Semantics and

See also the companion textbook *Style in Technical Math* by the author.

Semantics
and the Syntax of Algebra
Afshin Azari-Vala

Cover design by Afshin Azari-Vala

Published by Afshin Azari-Vala

To contact the publisher please send email to admin@tersemath.com.

ISBN 978-1-7750996-0-4

Disclaimer: Every effort has been made to make sure that this publication is free of error. However, the author and the publisher do not assume and hereby disclaim any liability to any party for any loss, damage, or disruption caused by any errors, typographical or other, that may have crept into this publication.

To Keon

Contents

Preface

This textbook and its companion, *Style in Technical Math*, present a novel approach to the teaching of the fundamentals of mathematics that is coherent, accessible and immediately applicable.[1] The main objective is to bring about fluency[2] in the use of mathematical tools in working with everyday life problems as well as a wide range of problems in pure and applied sciences.

To make sure that the presentation is coherent, we have imposed an overarching structure that connects the many seemingly disparate concepts, tools and techniques in mathematics into a single narrative. The underlying theme in this narrative is that mathematical notation and algorithms emerge naturally as a result of our increasingly more complex problem-solving needs. We hope that the reader will find this view sufficiently stimulating to spark her or his interest in finding out more and thus keep reading.

To make the subject matter accessible, we have adopted a concrete to abstract approach in our coverage of the material. Each section presents a selection of word problems whose solutions require the use of similar techniques. We start with the familiar and wade into the unfamiliar in stages to ensure continuity.

To make the tools and techniques applicable, we promote algorithms that are meaningful and efficient. Where possible, alternative approaches to the art of problem-solving are presented and circumstances under which the use of one or some other alternative is in order are listed. Particular attention is given to algorithms that are used by those who exhibit fluency in the use of math.

This brings us to the notion of style. By *style* we mean the particular selection of tools that one makes when solving a given problem. Style matters. Good style ensures that what one does is meaningful, i.e., that the

---

[1] The novelty is in the use of semantics to teach the subject matter at the level of fundamentals. Such use of semantics has already been made in the teaching of other subject matters notably in teaching the fundamentals of logic in the publication *Leblanc, Hugues & Wisdom, William. Deductive Logic. 2nd Edition. Boston: Allyn and Bacon Inc., 1976. ISBN 0-205-05496-X.*

[2] By *fluency* we mean the ability to naturally select and seamlessly use tools that are meaningful and efficient and to do so with confidence.

activity is in line with the way we naturally reason, and that the activity is efficient enough to keep the conversation going. It is this combination of meaningfulness and efficiency that makes a mathematical tool truly powerful.

The textbook caters to the needs of a wide variety of readers. It is recommended for adult learners who wish to strengthen their understanding of the fundamentals of math particularly those who intend to move on to the study of pure and applied sciences. It is written for parents who wish to have an understanding of the organization of the subject matter and its application to everyday life problems and problems in the sciences in order to help them teach the subject to their kids and to do so with confidence. It is suitable for high school students in their senior years and college and university students in their junior years. It provides a good reading for those active in the pure and applied sciences including scientists and engineers who wish to have a highly organized understanding of the subject matter at the level of fundamentals. And it is written for those active in the field of linguistics as it offers deep insight into the syntactic evolution of mathematical language to deal with our daily needs and the needs of the sciences.

The author has taught the fundamentals of mathematics to adult learners for over 25 years and, in addition to briefly pursuing studies in music, the sciences (physics and biology), and dentistry, holds bachelor's and master's degrees in aerospace engineering, with unfinished studies at the level of PhD. The author has taught courses in mathematics, physics, chemistry and computer programming at various institutions and colleges and currently teaches mathematics at George Brown College in Toronto, Canada. The present textbook and its companion, *Style in Technical Math*, advance the point of view of the author on the manner in which mathematical knowledge at the level of fundamentals should be organized and taught.

The author wishes to thank his current and former students for the many engaging discussions and questions that helped immensely in shaping this textbook. Words of gratitude are also due to current and former Deans, Ian Wigglesworth and Georgia Quartaro, and Chairs, Alex Irwin, Gerry Conrad, Susan Toews, and Tony Priolo, for their support, as well as the many professors who used this textbook or earlier versions of it in their courses and provided the author with valuable feedback. This includes professors Marie Jaffe, John Waters, Sorina Zota, Michael Matisko, Bartek Roszak, Nader Afrand, Negica Popovic, Rebecca Pali, Mazdak Nik-Bakht, Steven Konvalinka, Jeff McManus, Natalie Drumonde and Elisa Romeo.

The author is indebted to Prof. Jeff McManus in particular for the few intriguing weekly discussions at the cafe across the street. It was reflection on your objection to the association made in the textbook between direct proportion and multiplication during these discussions that led the author to the formulation of the standard and conservation forms of the models for the many problem types[3] which greatly improved the structure of the

---

[3]The terminology is borrowed from aerodynamics.

presentation on alternative models for the various problem types.

Special thanks are due to my lovely wife, Nooshin Mohtashami-Maali, for the love, patience and understanding over the past twenty years that I spent working on these pages as well as my son, Keon, to whom this work is dedicated, for livening up the red and the blue with green.

Introduction

Part I of this textbook presents an overview of the problem-solving process and introduces two problem-solving techniques, referred to in this textbook as the *step-by-step technique* and *the algebraic technique*. A number of problems are presented to illustrate both the difference in approach and the connection between the two techniques. A brief list of the advantages of the algebraic technique over its step-by-step counterpart is presented and explored in detail later in the text.

In the coverage of the manner in which algebraic solutions are set up, we pay particular attention to the key activity of modelling word problems by relating the values of the various quantities involved in the word problem and expressing this relationship as an algebraic equation. This naturally leads to a study of the syntax of algebra and the manner in which semantics are carried by this syntax.

In Parts II to V we present a systematic study of the various types of equations that are generated through the modelling stage in the algebraic technique. The coverage follows the level of complexity of the kinds of problems that we solve beginning with the simplest types of problems and moving on to more and more complex types of problems. This order beings with superposition problems (Part II), followed by proportion problems (Part III), followed by linear problems (Part IV), followed by nonlinear problems (Part V). A wide range of everyday life problems and problems in basic sciences are used to illustrate the manner in which algebraic solutions are deployed.

In Part VI we extend the ideas that have been developed so far to discuss the origin of formulas and the manner in which one extracts information from formulas by relating the values of the various quantities that appear in them, as well as the manner in which one can rearrange formulas.

Appendices are used to provide background information that is needed throughout the textbook (Appendix A on quantities, Appendix B on concepts from chemistry and Appendix C on grading schemes), as well as alternative models that are commonly used in solving proportion problems (Appendix D on alternative techniques for solving direct proportion problems and Appendix E on alternative techniques for solving inverse proportion problems).

Numerous problems are included in the exercises at the end of each section and subsection. The level of the problems ranges from simple to challenging. Answers and detailed solutions to the problems in the exercise sets are given in the companion solution manual.[4]

Ideally the textbook should be covered in tandem with the companion textbook *Style in Technical Math* by the author in the same course with equal times allocated to the coverage of each textbook. If both texts are used simultaneously, high school curricula and upgrading programs may opt to cover Chapters 1 to 4 in one term and the rest in another while the entire textbook may be covered in a single term at college and university level programs. If the companion text is not covered, the times recommended above may be cut in half.

---

[4]In order to bring about fluency in solving mathematical problems, one would have to have access to detailed solutions to the problems that are posed in the exercises in the textbook and not just the final answer. For this reason, we have not provided the answers to the problems in the exercises at the end of the textbook and since appending full solutions to the problems would require the addition of many pages, it was decided to place the solutions in a separate document.

Part I

Problem-Solving
Fundamentals

In this introductory part we focus on the process of problem-solving itself. Following a discussion on word problem structure, we present two problem-solving techniques in common use and apply them to solve a selection of problems whose solutions require the use of the four basic arithmetic operations. Along the way we discuss terminology and notation, and set a few conventions that facilitate the task of communicating the solution.

Chapter 1
The Problem-Solving Process

1.1   The Structure of a Word Problem

A problem whose solution requires the use of math is often stated in prose and, as such, is referred to as a *word problem*. Such statements are often organized into two sections. The first provides information on the values of certain quantities that can be used to calculate the values of one or more other quantities.[1] The second specifies what these other quantities are and asks the reader to calculate their values in certain units using the given information. This request is usually phrased as an imperative or an interrogative statement.[2]

Example

first section — Since noon, Paula had 2 cups of tea at 24 Cal/cup, 3 slices of pizza at 181 Cal/slice and a slice of cake containing 235 Cal. She was inactive for 1.5 h, burning, on average, 80 Cal/h. Later, she went running and burnt 340 Cal.

second section — Calculate Paula's net intake of energy since noon in Calories.

The first section in the composition of the sample word problem above provides us with information on the values of the quantities *number of cups*

---

[1] It should be immediately apparent that an understanding of what quantities are is essential when working with word problems. For a detailed discussion of quantities, their names, symbols and values please see the chapter on measurement in the companion textbook *Style in Technical Math* by the author. A condensed version of the treatment in the companion textbook is given in Appendix A for the reader's convenience.

[2] It is usually not difficult to locate such statements. Key words and phrases to look for include such words and phrases as *calculate*, *find*, *what is* and the like.

of tea, energy content per cup of tea, number of slices of pizza, etc., that can be used to calculate the value of some quantity of interest. The second section in the composition of the sample word problem above tells us that the quantity of interest, i.e., the quantity whose value we seek, is *energy*, specifically, *Paula's net intake of energy since noon* and it asks us to use the unit Calorie in the calculation of its value.

1.1.1   Variations in Structure

The two sections in the composition of the sample word problem above are quite distinct and the reader should be aware that variations do exist. For one thing, as noted earlier, the request to calculate the values of the quantities of interest may be made through an interrogative, as opposed to imperative, statement as in

What is Paula's net intake of energy since noon in Calories?

In addition, often further information is bundled into the second section within the imperative/interrogative statement as in

The half-life of a radioisotope is 3.2 days. What mass of the radioisotope will be left after 7.1 days if we start with 20.8 g of the radioisotope?

And from time to time statements providing more information follow the imperative/interrogative statement as in

The potential energy of a 1200 kg car is 31 000 J. Find the altitude at which the car is located. Acceleration due to gravity near the surface of the Earth is approximately 9.8 m/s^2.

1.1.2   References to Quantities Whose Values Are Sought

The wording of the problem may not explicitly name the quantities whose values are sought as in

You started with 25.8 mL of a salt solution and you have 12.8 mL of the solution left. How much salt solution did you use?

The interrogative statement in the composition above uses the more general phrase *how much* in place of the more specific phrase *what volume of*. The argument is that the use of the unit *millilitre* in the expressions of the initial and final values of the volume of the salt solution, and the fact that the value of the quantity whose value is sought is the difference between the final and initial values of the volume of the salt solution, imply that the quantity whose value is sought is one of volume. However, it still remains true that the use of such general phrases affects clarity especially as the level of complexity of problems rises.[3]

In some settings it is also common to use the unit name in place of the quantity name as in

> On average we can extract 4 Cal/g of carbohydrate, 9 Cal/g of fat, and 4 Cal/g of protein. A cup of yogurt contains 12.5 g of carbohydrate, 6.8 g of fat, and 11.1 g of protein. What is the cup of yogurt's Calorie content?

Note that the interrogative statement asks for the cup of yogurt's *Calorie* content in place of the more accurate *energy* content.

1.1.3 Specification of Units

In some cases the units to be used in the evaluation of the values of the quantities of interest are stated explicitly as in

> Calculate the mass of a molecule of H_2O in amu. The atomic masses of H and O are 1.008 amu/atom and 16.00 amu/atom respectively.

In other cases the choice of units in calculating the values of the quantities of interest may be set by convention as in the sciences where the units introduced by the International System of Units, SI, are used by default. In the following problem from physics

> A 3.82 kg object is moving at a speed of 0.22 m/s. What is the object's kinetic energy?

it is expected that the value of the quantity *energy* will be calculated in joules, the SI unit of energy. Adoption of a single set of units facilitates communication and eases the task of data comparison across disciplines.[4]

---

[3] In this textbook we will remain formal and always name the quantity whose value is sought. This will provide the reader with a formal structure against which she or he can gauge other constructs.

[4] Unfortunately, the convention on the use of SI units in the expression of the values of the quantities of interest in the sciences is not followed by all scientists or those who specialize

The choice of the units in the calculation of the values of the quantities of interest may also be implied as in

> You are travelling at a speed of 120 km/h. What distance will you cover in 2.5 h?

In such cases one must rely on other information given in the body of the word problem to determine what units one must use in the calculation of the values of the quantities whose values are sought. In the problem above, the value of the quantity *speed* is given in units of km/h and the value of the quantity *time* is given in units of h. It is, therefore, not unreasonable to conclude that the value of the quantity *distance* should be calculated in units of km.

No matter what variation is adopted, at some point the word problem asks the reader to calculate the values of one or more quantities in specified, agreed-upon, or implied units.[5] It is expected that the reader has the skill to analyze the word problem to determine and formally state what these quantities, and the units that should be used in the calculation of their values, are.

We end this section with a word problem that asks for the calculation of the values of multiple quantities.

> An order of 40 pens and 8 binders costs $135.52. An order of 35 pens and 9 binders costs $132.56. Calculate the cost of a pen and the cost of a binder.

Exercise Set 1.1

For each problem, identify the quantities whose values are sought and the units in which their values should be evaluated. Quote quantity names, their associated quantity symbols, unit names and their associated unit symbols.

1. A boy pulls a block over a distance of 4.1 m with a force of 12 N acting at an angle of 30° upward from the horizontal. Determine the work done by the boy.

2. $\frac{2}{3}$ of the workers will be laid off next week. How many workers will be laid off next week if, at present, there are 542 workers?

---

in applied science fields. Physicists tend to follow this convention more than chemists and life scientists. As for applied sciences, many have their own (separate or additional) set of units either because such non-SI units have features that make them suitable for use in those particular disciplines or else because they have persisted historically.

[5] If doubt persists, you should request clarification.

3. A voltage difference of 120 V across a wire sets up a current of 0.3 A in the wire. What is the wire's resistance?

4. 3 half-lives have passed since 0.8 g of I-131, a radioisotope, was generated. What mass of I-131 in grams is left?

5. Consider the following chemical reaction taking place under STP conditions:

$$C_3H_8 + 5O_2 \rightleftharpoons 3CO_2 + 4H_2O$$

What volume of CO_2 in litres is generated by the complete combustion of 120 g of C_3H_8?

6. A 2.5 kg particle is moving around a circle of radius 2.3 m at a speed of 0.93 m/s. What centripetal force acts of the particle?

7. Statistics show that, in an outbreak of a certain virus in a community, 11.5% of the people in the community contract the virus. Of these, 61.4% seek medical help. If there is an outbreak in a community of 4000 people, how many people are expected to contract the virus but not seek medical help?

8. In a sample of frogs, 8.5% have red eyes. Of these, only 22% survive after two weeks. How many red-eyed frogs will survive after two weeks if we start with a sample of 308 frogs?

9. A hat is on sale at 20% off. Tax is added at 13%. Find the amount of tax if the amount of discount is $2.10.

10. We have 32.5 mol of helium in a container of volume 0.5 m^3. Calculate the pressure of the helium if its temperature is 290 K. Use 8.314 J/(mol·K) as the value of the ideal gas constant.

11. 210 eligible voters voted. This represents 62% of the eligible voters. How many eligible voters are there?

12. I bought 3 pens at $2.20/pen, 4 notebooks at $3.99/notebook and a ruler for $4.25. How much did I spend in total?

13. The kinetic energy of a particle is 0.18 J. Find the speed of the particle if it has a mass of 0.25 kg.

14. A rod, made of steel, has a temperature of 32 °C and a length of 12.6 cm. Find the length of the rod if its temperature changes to 44 °C. The coefficient of linear thermal expansion for steel is 1.30×10^{-5} 1/°C.

15. A book, priced at $12.99, is on sale. Calculate the rate of discount if the sale price is $9.74.

16. Calculate the mass of a molecule of CO_2 in amu. The atomic masses of C and O are 12.01 amu/atom and 16.00 amu/atom respectively.

17. The patient has a temperature of 102.2 °F. What is the temperature of the patient in °C?

18. A garden has the shape of a trapezoid with bases that measure 4.1 m and 3.2 m and an altitude that measures 3.7 m. Calculate the area of the garden.

1.2   Problem-Solving Techniques

The process of calculating the values of the quantities that a word problem asks for using the values of the quantities that it provides is called *solving the word problem*. In this context, the word *answer* is used to refer to a list of the calculated values of the quantities that the word problem asks for and the word *solution* is used to refer to a detailed presentation of the manner in which the problem is solved.

In this section we will present two techniques that are commonly used to solve word problems. We refer to these as the *step-by-step technique* and the *algebraic technique*. The two differ greatly in their approach to the art of solving word problems.

1.2.1   The Step-by-Step Technique

Using the *step-by-step technique*, one uses logic to list a set of steps that can be taken to calculate the values of the quantities that the word problem asks for using the values of the quantities that it supplies. The solution to the word problem is then communicated by providing, for each step, a brief note on its objective in prose, followed by a presentation of any associated calculations and a closing remark on the result of these calculations.

For the sake of uniformity, in this textbook we will follow the conventions listed below to communicate solutions that employ the step-by-step technique.

1. Steps and their recursive substeps are labelled using numbers, lowercase letters and lowercase Roman numerals.

2. Within each step or recursive sub-step, we
   a. state the aim of that step or sub-step in prose,
   b. perform any associated calculations, and
   c. discuss the result in prose.

Technical tools are used to perform the calculations at each step or substep. The question of how one differentiates between a step and a sub-step will be addressed in the pages ahead.

To illustrate how the step-by-step technique works, let us return to the sample word problem that was posed at the beginning of this chapter. One way to solve this problem is to take the steps listed below.

1. Calculate Paula's intake of energy from the cups of tea.
2. Calculate Paula's intake of energy from the slices of pizza.
3. Calculate Paula's loss of energy due to inactivity.
4. Calculate Paula's net intake of energy since noon.

We communicate the full solution as follows:

A Step-by-Step Solution

1. Calculate Paula's intake of energy from the cups of tea.

$$24 \times 2 = 48$$

Paula had an intake of energy of 48 Cal from the cups of tea.

2. Calculate Paula's intake of energy from the slices of pizza.

$$181 \times 3 = 543$$

Paula had an intake of energy of 543 Cal from the slices of pizza.

3. Calculate Paula's loss of energy due to inactivity.

$$80 \times 1.5 = 120$$

Paula had a loss of energy of 120 Cal while she was inactive.

4. Calculate Paula's net intake of energy since noon.

$$48 + 543 + 235 - 120 - 340 = 366$$

Since noon, Paula had a net intake of energy of 366 Cal.

Exercise Set 1.2.1

Solve each of the following word problems using the step-by-step technique.

1. You bought 3 pens at \$4.20/pen and 5 notebooks at \$6.99/notebook. What amount of money did you spend in total?

2. Used textbooks are sold at $42.50/textbook and new ones are sold at $75.99/textbook. What amount of money does the bookstore receive from the sale of 97 used textbooks and 54 new textbooks?

3. On average we can extract 4 Cal/g of carbohydrate, 9 Cal/g of fat, and 4 Cal/g of protein. A cup of yogurt contains 6.2 g of carbohydrate, 8.5 g of fat, and 13.7 g of protein. Calculate the energy content of the cup of yogurt in Cal.

4. This afternoon Aisha had 3 cups of coffee at 30 Cal/cup, 2 servings of fruit at 110 Cal/serving and toast and butter containing 87 Cal. She then worked at her desk for 2 h, burning, on average, 130 Cal/h. Calculate Aisha's net intake of energy this afternoon.

5. It costs $2.50 to mail a letter, $8.10 to mail a package, and $19.50 to mail a box. What is the cost of mailing 18 letters, 12 packages, and 8 boxes?

6. This week Joan worked 8 hours as a clerk earning $16.35/h, 4.5 hours as a waitress earning $15.20/h, and 12 hours at a fundraising event earning $12.50/h. What is Joan's total earnings this week?

7. Calculate the mass of a molecule of N_2O_3. The atomic masses of N and O are 14.01 amu/atom and 16.00 amu/atom respectively.[6]

8. This morning you drove at 110 km/h for 2.5 h. This afternoon you drove at 90 km/h for 1.5 h. What total distance did you cover today?

9. Over the past 2.75 h recycling material has been arriving at the plant at a rate of 17.5 t/h. During the last 1.5 h, the plant has been processing the recycling material at a rate of 14 t/h. What mass of recycling material is still waiting to be processed?

10. Your goal is to cycle a distance of 100 km this week. Over the past three days, you cycled at 20 km/h for 0.75 h, 18.5 km/h for 1.5 h, and 23.5 km/h for 0.5 h. What additional distance do you have to cover this week to reach your goal?

1.2.2   The Algebraic Technique

An alternative to the step-by-step technique is to kick-start the solution process by relating the values of the various quantities in the statement of the word problem in the simplest possible manner and then process this relationship until we arrive at a value for each of the quantities whose value is sought by the word problem. The relationship between the values of the various quantities involved, called a **model**, can be expressed using the language of algebra and often takes the form of an equation. We refer to this approach

---

[6]For a brief coverage of basic concepts in chemistry please see Appendix B.

as the *algebraic technique.*

Three distinct stages can be identified in the algebraic technique. These are

1. Model: Establish the simplest possible relationship between the various quantities in the statement of the word problem and write this relationship as an algebraic equation.

2. Solve: Use semantic tools to logically manipulate the algebraic equation until the values of the quantities whose values are sought by the word problem are determined.

3. Interpret: Make sense of the algebraic solution in the context of the word problem.

Figure 1.2.1 illustrates the stages in the algebraic technique.

To illustrate matters, we present an algebraic solution to the sample word problem posed at the beginning of this chapter.

An Algebraic Solution for the Sample Word Problem

E Paula's net intake of energy since noon (Cal)
$$E = 24 \times 2 + 181 \times 3 + 235 - 80 \times 1.5 - 340$$

$$E = 48 + 543 + 235 - 120 - 340$$
$$E = 366$$

Since noon, Paula had a net intake of energy of 366 Cal.

The stages in the algebraic solution to the sample word problem above are shown below.

Stage 1 Model

E Paula's net intake of energy since noon (Cal)
$$E = 24 \times 2 + 181 \times 3 + 235 - 80 \times 1.5 - 340$$

Stage 2 Solve

$$E = 24 \times 2 + 181 \times 3 + 235 - 80 \times 1.5 - 340$$
$$E = 48 + 543 + 235 - 120 - 340$$
$$E = 366$$

Stage 3 Interpret

Since noon, Paula had a net intake of energy of 366 Cal.

```
┌─────────────────────────────────────────┐
│       Statement of the word problem       │
│       involving the various quantities    │
└─────────────────────────────────────────┘
```

Model: Establish the simplest possible
relationship between the values of the
various quantities in the statement
of the word problem

```
┌─────────────────────────────────────────┐
│   Expression of the relationship between  │
│   the values of the various quantities in │
│   the statement of the word problem as    │
│          an algebraic equation            │
└─────────────────────────────────────────┘
```

Solve: Use semantic tools to logically
manipulate the algebraic equation until you
arrive at the values of the quantities whose
values are sought by the word problem

```
┌─────────────────────────────────────────┐
│      Solution to the algebraic equation   │
└─────────────────────────────────────────┘
```

Interpret: Make sense of the
algebraic solution in the context
of the word problem

```
┌─────────────────────────────────────────┐
│          Solution stated in English       │
└─────────────────────────────────────────┘
```

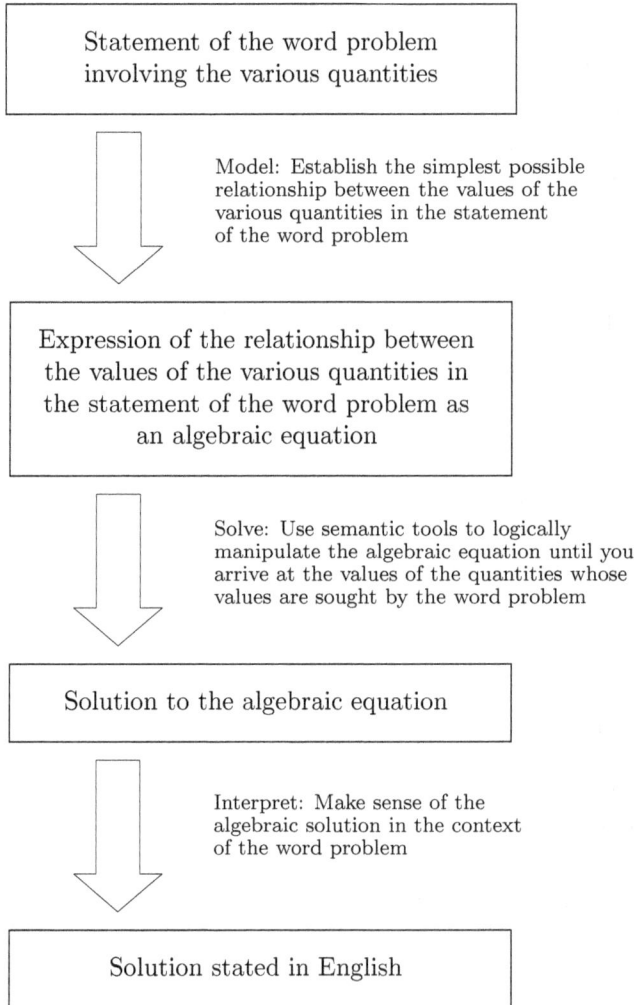

Figure 1.2.1: Stages in the algebraic technique

We will now discuss the three stages in the algebraic solution above in more detail.

Stage 1: Model

This stage generates two lines.[7] For the sample word problem above, the two lines are reproduced below.

$$E \quad \text{Paula's net intake of energy since noon (Cal)}$$
$$E = 24 \times 2 + 181 \times 3 + 235 - 80 \times 1.5 - 340$$

The first line is provided for convenience as a quick reference. It lists the symbol for the quantity whose value is sought by the word problem, followed by a phrase describing the quantity whose value it represents, followed in turn by the unit symbol used to express its value within parentheses.[8] This is shown in Figure 1.2.2 below.

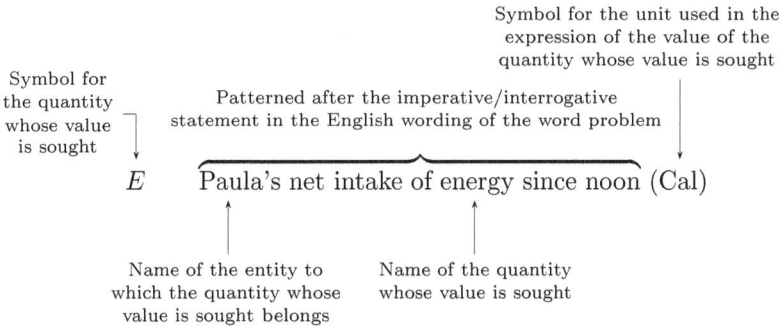

Figure 1.2.2: Analysis of the structure of the first line of the model in the solution of the sample word problem

The choice and casing of the symbol used to represent the value of the quantity whose value is sought is important. Unless conventions in your field

---

[7]In general, it generates two *sections*. The first section lists the symbol for each quantity whose value is sought by the word problem, followed by a phrase describing the quantity whose value it represents, followed in turn by the unit symbol used in the expression of its value within parentheses. The second section lists the constraints that the word problem imposes on the values of the various quantities.

[8]For a detailed discussion of quantities, their names, symbols, and values please see the chapter on measurement in the companion textbook *Style in Technical Math* by the author. A condensed version of the treatment in the companion textbook is given in Appendix A for the reader's convenience. The reader should read this appendix now if she or he has not done so already.

dictate otherwise, choose the first letter of the name of the quantity whose value is sought as the quantity symbol representing the value of that quantity and feel free to use whatever casing that suits your taste but, once you have made a decision, stick to that casing throughout the solution.[9] In the example above, the quantity whose value is sought is *energy*, hence the choice of the letter E as the quantity symbol representing its value. The choice of the letter and its casing are both conventions in the sciences.

To decide how to phrase the statement that describes the quantity whose value is sought, look at how the imperative/interrogative statement in the English wording of the word problem is put together and use this *as a guide*.[10] Make sure to name the quantity whose value is sought and the entity to which the quantity belongs.

Following this statement write the unit symbol (not the unit name) used in the expression of the value of the quantity whose value is sought within parentheses.

The second line in the modelling stage presents a statement which relates the values of the various quantities involved in the statement of the word problem in the simplest possible manner. As stated earlier, this relationship is expressed using the language of algebra and often takes the form of an equation.[11] For the sample word problem above, we have chosen to relate the values of the various quantities involved by noting that Paula's net intake of energy since noon can be obtained by adding the intakes of energy from the tea, pizza and cake and subtracting the losses of energy due to inactivity and running.[12] The relationship between the values of the various quantities stated above can be written using the language of algebra as follows.[13]

$$E = 24 \times 2 + 181 \times 3 + 235 - 80 \times 1.5 - 340$$

---

[9]This last requirement is important. If both the lowercase and uppercase versions of the same letter are present in a solution, it is understood that they refer to the values of *different* quantities.

[10]This does not mean that you should take the imperative/interrogative statement in the modelling stage and use it verbatim to describe the quantity whose value is sought. Such an approach usually fails to meet the standards of English grammar.

[11]An **expression** is a meaningful sequence of numbers, quantity symbols and operations other than order operations such as the equality operation. An **equation** consists of two expressions that are set equal to each other.

[12]This is not the only way to relate the values of the various quantities involved in the statement of the word problem. An alternative relationship would be *the sum of the values of energy intakes should equal to the sum of the value of the stored energy and values of energy losses, i.e.,*

$$24 \times 2 + 181 \times 3 + 235 = E + 80 \times 1.5 + 340$$

Of the many possible alternatives, those that are the simplest are given preference as they are easy to remember and recall. We will return to this topic later in the text.

[13]Note that all units are dropped at this point.

To see how the equation above relates the values of the various quantities involved as intended, you must train yourself to break it down into pieces as shown below.

$$\boxed{E} \;=\; \boxed{24 \times 2} \;+\; \boxed{181 \times 3} \;+\; \boxed{235} \;-\; \boxed{80 \times 1.5} \;-\; \boxed{340}$$

Each box in the illustration above is called a **term**. On the left side of the equation, i.e., the left side of the equality sign, we have one term which contains the symbol for the quantity whose value is sought. On the right side of the equation, i.e., the right side of the equality sign, we use addition symbols and subtraction symbols as a guide to break the expression into terms.[14]

The ability to break an algebraic expression into terms is as fundamental as the ability to break an English sentence into words. And just as words have meanings, so do terms. In the equation above, the term on the left represents Paula's net intake of energy since noon (as specified by the first line of the model). On the right side of the equation, the first term on the left represents Paula's net intake of energy from the tea, the second term from the left represents Paula's net intake of energy from the consumption of the slices of pizza, the third term from the left represents Paula's intake of energy from the consumption of the slice of cake, the fourth term from the left represents Paula's net loss of energy due to inactivity, and the rightmost term represents Paula's net loss of energy due to running.[15] One can write the relationship between the values of the various quantities involved as an equation using the language of algebra, but one can also revert back to the use of English to state the relationship so expressed as shown below.

---

[14]The reader should note that we use boxes for illustration purposes only. One should not develop a dependence on the presence of boxes or other visual cues such as ovals, underlined terms, etc., to identify the terms in an expression. In particular, you should steer clear of the use of parentheses for the identification of terms as their use in this context conflicts with another convention, introduced later, that limits the use of brackets to imply multiplication. What we do recommend is the use of empty spaces on the left and right sides of each + and each − to help set the terms apart visually.

[15]Note how each phrase describing a term names the relevant quantity and the entity to which it belongs.

	is equal to	plus		plus		minus		minus	
$\boxed{E}$	$=$	$\boxed{24 \times 2}$	$+$	$\boxed{181 \times 3}$	$+$	$\boxed{235}$	$-$	$\boxed{80 \times 1.5}$	$-$	$\boxed{340}$
Paula's net intake of energy since noon in Calories		Paula's intake of energy from the tea in Calories		Paula's intake of energy from the pizza in Calories		Paula's intake of energy from the cake in Calories		Paula's loss of energy due to inactivity in Calories		Paula's loss of energy due to running in Calories

Note that the English translation of the algebraic statement does not refer to actual numerical values but names the quantities whose values are related in the manner that they are.

Stage 2: Solve

The steps at this stage are reproduced below.[16]

$$E = 24 \times 2 + 181 \times 3 + 235 - 80 \times 1.5 - 340$$
$$E = 48 + 543 + 235 - 120 - 340$$
$$E = 366$$

We now discuss the steps in detail.

The understanding that an algebraic expression breaks into terms helps us understand how to evaluate such expressions: Evaluate each term and then add and subtract the terms. For the example above, evaluation of the terms yields the following.

$$\boxed{E} = \boxed{24 \times 2} + \boxed{181 \times 3} + \boxed{235} - \boxed{80 \times 1.5} - \boxed{340}$$

$$\boxed{E} = \boxed{48} + \boxed{543} + \boxed{235} - \boxed{120} - \boxed{340}$$

It is important that the reader should train herself or himself to pause once the terms in an expression are evaluated and extract information from the values of the various terms before proceeding further. A quick scan of the values of the various terms involved in the equation above would immediately

---

[16]The first line is a repeat of the model from Stage 1 and does not need to be listed again; hence, its faded appearance. We have included the line here to remind the reader that Stage 2 begins with the model that is established in Stage 1.

tell us that, of the various sources of intake, the intake of energy from the consumption of the slices of pizza was the largest, much larger than the intake of energy from the cups of tea and more than twice the intake of energy from the consumption of the slice of cake. In addition, we can also see that most of the loss of energy resulted from running, and, while the loss of energy due to running did not offset the intake of energy from the consumption of the slices of pizza, it did more than compensate for the gain of energy from the consumption of the slice of cake.

Having compared the sizes of the various terms in the context of the word problem above, we now proceed to add and subtract the terms to evaluate Paula's net intake of energy since noon. Here, we adopt the view that *each operation applies to the term that follows it*. This is illustrated below.

$$48 \; + \; 543 \; + \; 235 \; - \; 120 \; - \; 340$$

As you can see, the term 543 is being added. The term 235 is also being added. The term 120, on the other hand, is being subtracted as is the term 340. The term 48 on the left end is also being added as its sign is positive.[17]

To evaluate the expression on the right, then, we start with 48 and work our way to the right, adding and subtracting the terms as indicated,[18] i.e., we start with 48, add 543 to it, take the result and add 235 to it, take this result and subtract 120 from it, and take the latest result and subtract 340

_____

[17]The term is being added to 0, i.e., we see this as $0 + 48$. For an in-depth discussion of positive and negative signs and their relation to addition and subtraction symbols please see the chapter on *the formal classification scheme* in the companion textbook *Style in Technical Math* by the author.

[18]We could, in fact, apply the operations in any order that we please so long as we keep in mind that each operation applies to the term that follows it. We could, as an example, start with 235, subtract 120 from it, add 48 to the result, add 543 to this result, and subtract 340 from the latest result, i.e.,

$$235 \; - \; 120 \; + \; 48 \; + \; 543 \; - \; 340$$

and we would still arrive at the correct answer. Another equivalent order would start with a loss of 120, followed by a gain of 48, followed by a loss of 340, followed by a gain of 235, followed by a gain of 543, i.e.,

$$-120 \; + \; 48 \; - \; 340 \; + \; 235 \; + \; 543$$

All of this should be somewhat obvious in the context of the sample word problem above as it should not matter whether we add, say, the gain of 543 Cal, sooner or later so long as we do so at some point. The order in which the terms are listed in the algebraic solution in the main body of the text above is, however, preferred on other grounds: The order of the listing of the terms maps onto the order in which the values are introduced in the body of the word problem. This makes it easier for the reader to go back and forth from the equation to the wording of the problem and make matches.

from it. This evaluates to 366 so that we can write[19]

$$E \;=\; 366$$

In summary, to evaluate an expression involving $+$, $-$ and $\times$, evaluate the terms in one step, and then add and subtract them at the next step.[20]

---

[19] As you go from line to line, carry the whole equation not just the expression on the right side. The solution

$$\begin{aligned} E \;&=\; 24 \times 2 \;+\; 181 \times 3 \;+\; 235 \;-\; 80 \times 1.5 \;-\; 340 \\ &=\; 48 \;+\; 543 \;+\; 235 \;-\; 120 \;-\; 340 \\ &=\; 366 \end{aligned}$$

is *not* recommended. While there is nothing wrong with this shorthand presentation, as our aim is to set an introduction to algebra proper where the unknown may appear anywhere in the equation – at which time we will have no option but to work with the equation as a whole as we go from one line to the next – we prefer to adopt the same presentation style here to make the transition seamless, i.e., we prefer the presentation

$$\begin{aligned} E \;&=\; 24 \times 2 \;+\; 181 \times 3 \;+\; 235 \;-\; 80 \times 1.5 \;-\; 340 \\ E \;&=\; 48 \;+\; 543 \;+\; 235 \;-\; 120 \;-\; 340 \\ E \;&=\; 366 \end{aligned}$$

[20] The algorithm proposed in the text above is called the *analysis-synthesis algorithm* and consists of two stages: During the first stage, called *analysis*, we break the given expression into terms (entities within the expression that are being added or subtracted), and the terms into **factors** (entities within terms that are being multiplied). During the second stage, called *synthesis*, we start with the factors, work them out, multiply them to work out the terms, and then add and subtract the terms to evaluate the expression.

A different algorithm in common use is the *step-by-step BEDMAS algorithm* which instructs one to start from the left and repeatedly work out the brackets, then exponents, then divisions, then multiplications, then additions and finally subtractions. The use of this algorithm in the solution of the sample word problem above would lead to the following lines:

$$\begin{aligned} E \;&=\; 24 \times 2 \;+\; 181 \times 3 \;+\; 235 \;-\; 80 \times 1.5 \;-\; 340 \\ E \;&=\; 48 \;+\; 181 \times 3 \;+\; 235 \;-\; 80 \times 1.5 \;-\; 340 \\ E \;&=\; 48 \;+\; 543 \;+\; 235 \;-\; 80 \times 1.5 \;-\; 340 \\ E \;&=\; 48 \;+\; 543 \;+\; 235 \;-\; 120 \;-\; 340 \\ E \;&=\; 591 \;+\; 235 \;-\; 120 \;-\; 340 \\ E \;&=\; 826 \;-\; 120 \;-\; 340 \\ E \;&=\; 706 \;-\; 340 \\ E \;&=\; 366 \end{aligned}$$

There are numerous advantages to the use of the analysis-synthesis algorithm over the step-by-step BEDMAS algorithm when evaluating expressions such as efficiency, ease of trouble-shooting, and the like. In the context of solving word problems, however, the most important advantage of the analysis-synthesis algorithm over the step-by-step BEDMAS algorithm is that the analysis-synthesis algorithm is in line with the manner in which we reason when we solve problems. As an example, in the context of the word problem above, it is quite clear why we need to carry out the multiplications first before we add and subtract, i.e.,

$$\begin{aligned} E \;&=\; 24 \times 2 \;+\; 181 \times 3 \;+\; 235 \;-\; 80 \times 1.5 \;-\; 340 \\ E \;&=\; 48 \;+\; 543 \;+\; 235 \;-\; 120 \;-\; 340 \\ E \;&=\; 366 \end{aligned}$$

The reason, of course, is that one needs to first work out the net gains and net losses due to various sources and then add and subtract these to compute the overall net gain. The

Stage 3: Interpret

At this stage we make sense of the solution to the equation in the context of the word problem. We present this as a full sentence that answers the question posed by the word problem. For the sample word problem above, this line is

Since noon, Paula had a net intake of energy of 366 Cal.

You can use the statement that you wrote during the modelling stage to describe the quantity symbol for the quantity whose value is sought as a guide to compose the closing statement at the interpretation stage. Make sure to name the quantity whose value is sought as well as the entity to which that quantity belongs.

Note also that the unit symbol is put back in at this stage.

Notes on Formulating a Model

When setting up the equation that models a problem, one may use any valid relationship between the values of the quantities involved in that problem to base the equation on. As an example, consider the following problem.

Problem

Davida had $54.20. She bought a pair of gloves for $8.99 and a hat for $22.50. How much money does Davida have left?

One way to relate the various quantities involved in this problem is to subtract, one by one, the amounts of money spent from the amount of money Davida started with. Such a relationship leads to the following equation as the model for the problem.

a amount of money Davida has left ($)
$a = 54.20 - 8.99 - 22.50$

A different line of reasoning would argue that amount of money left is equal to amount at the start minus the sum of all expenses. This line of

---

step-by-step BEDMAS algorithm, on the other hand, provides a set of steps that should be followed for no apparent reason other than that we are assured by trusted authorities that they work. For more on the advantages of the use of the analysis-synthesis algorithm over the step-by-step BEDMAS algorithm please see the companion textbook *Style in Technical Math* by the author.

reasoning leads to the following equation as the model for the problem.

> a amount of money Davida has left ($)
> $a = 54.20 - (8.99 + 22.50)$

Both equations are valid models for the problem above and there are others. However, note that, while any valid relationship may be used in the formulation of the equation that models the problem, some formulations are easier to remember as the relationship that they express is simpler. In general, formulations that are based on simpler relationships are preferred and are the ones used by most.

When formulating your equation, make sure that any values used in the equation can be found in the statement of the word problem otherwise the solution is not algebraic but step-by-step. As an example, one may use the model

> a amount of money Davida has left ($)
> $a = 54.20 - (8.99 + 22.50)$

for the problem posed above as the values used in the equation, i.e., 54.20, 8.99 and 22.50 are all found in the body of the word problem. However, the following model

> a amount of money Davida has left ($)
> $a = 54.20 - 31.49$

is not acceptable as it uses the value 31.49 in its formulation but this value is not present in the statement of the word problem. Note that the value 31.49 is equal to the sum of the expenses 8.99 and 22.50. While the equation above is logically correct, it is not an algebraic solution but a step-by-step solution. The steps are

1. Calculate the sum of the expenses.

$$8.99 + 22.50 = 31.49$$

In total Davida spent $31.49.

2. Calculate amount of money left.

$$54.20 - 31.49 = 22.71$$

Davida has $22.71 left.

The step-by-step solution above is disguised as an algebraic solution by skipping the first step and introducing a quantity symbol, a, in the second.

Finally, all the values that are used in the solution process must appear in the equation that models the problem, i.e., the very first equation that you write, otherwise the solution is not algebraic but step-by-step. Here is an example.

a amount of money Davida has left ($)
$a = 54.20 - 8.99$

$a = 45.21$
$a = 45.21 - 22.50$
$a = 22.71$

This solution can be broken into two steps:

1. Calculate amount of money left after the purchase of the gloves.

$a = 54.20 - 8.99$
$a = 45.21$

After the purchase of the gloves, Davida has $45.21 left.

2. Calculate amount of money left after the purchase of the hat.

$a = 45.21 - 22.50$
$a = 22.71$

We should also point out that in both versions the use of the symbol a to represent different quantities is problematic. In the solution below

a amount of money Davida has left ($)
$a = 54.20 - 8.99$

$a = 45.21$
$a = 45.21 - 22.50$
$a = 22.71$

the line

a amount of money Davida has left ($)

sets the quantity symbol a to represent the value of the quantity *amount of money left after the purchases* whereas on the next line

$a = 54.20 - 8.99$

the quantity symbol a evaluates to a different quantity, namely *amount of money left after the purchase of the gloves*. The next line

$a = 45.21$

continues to set the quantity symbol a as *amount of money left after the purchase of the gloves* but the lines following it, i.e.,

$$a = 45.21 - 22.50$$
$$a = 22.71$$

both revert the setting of the quantity symbol a back to *amount of money left after purchases.*

One may not use the same symbol to represent two different quantities in a given problem.[21]

Exercise Set 1.2.2

Solve the problems in Exercise Set 1.2.1 using the algebraic technique.

1.2.3   Comparison of the Step-by-Step and Algebraic Techniques

Both techniques can be used to solve any math problem no matter how complex the problem may be.

The step-by-step technique is used quite frequently when solving the simpler, more familiar problems that we encounter on a daily basis and, therefore, a knowledge of this technique is necessary in order to communicate effectively with the majority who use it. The approach is also commonly used in introductory science textbooks, especially for audiences with basic algebraic skills.

There are numerous advantages to the use of the algebraic technique, however. Some are obvious. As examples, algebraic presentations are more efficient than step-by-step presentations, an advantage that becomes more and more relevant as the level of complexity of the problems increases; they are easier to comprehend and troubleshoot as they display the full logic to the solution on a single line; etc. But there are deeper advantages to the use of the algebraic technique. These other key advantages are listed below and will be revisited at various points later in the textbook.

1. The algebraic technique facilitates analysis and synthesis.[22] Analysis begins with the splitting of an expression into terms (entities that are

---

[21] For problems where multiple quantity symbols of the same kind are needed, subscripts are used to differentiate between them. As an example, in a problem where both the value of the mass of an electron and the value of the mass of a proton need to be represented, we can use m_e to represent the former and m_p to represent the latter. Note that in all cases the main symbol is the quantity symbol. The subscript acts as a qualifier.

[22] *Analysis* is the process of breaking a complex entity into its logical components and those components into theirs and so on until we arrive at the fundamental building blocks

being added and/or subtracted) and the breakup of the terms into factors (entities within terms that are being multiplied). By superimposing semantics on such structural elements, we can make sense of what is happening at the level of factors, terms, and, of course, the expression itself.

2. The algebraic technique enables us to kick-start the solution to word problems by relating the values of the various quantities involved in the simplest possible manner. This is a powerful tool as it provides a simple answer to one of the most common questions that problem-solvers have: Where do I start?

3. The algebraic technique helps us see similarities in equations that model highly disparate systems which, in turn, helps us classify problems according to the types of operations that we see in the equations that model them. This allows us to deploy solution techniques based on problem type regardless of the context in which such problems arise.

4. The algebraic technique helps one gain a deeper understanding of formulas and fosters the ability to see relationships between the values of the various quantities involved in the making of a formula.

5. The algebraic technique makes it easier to impose the rules of rounding in the problem-solving process. Such impositions on algebraic presentations usually require fewer applications of the rules through the visual grouping of the values involved into factors, terms and the ultimate expression.

6. The algebraic technique can be used as a powerful self-teaching tool, allowing us to answer many questions through the manipulation of structural elements in equations that model our problems and extract step-by-step solutions to problem whose algebraic solutions are known.

It is difficult not to be impressed by the power and beauty of the algebraic technique. But perhaps the most pleasant reward of mastery of use of the algebraic technique in solving word problems is in the degree of confidence that it generates in those who use it in a meaningful way.

---

that figure in the composition of the complex entity with the aim of developing a deeper understanding of the nature of that entity. *Synthesis* is the process of assembling the building blocks that figure in the composition of an entity with the aim of rebuilding that entity.

Chapter 2
Terminology, Notation and Conventions

In this chapter we will discuss terminology and notation, and list conventions that ease the task of communicating the solution to word problems. We will do so through the study of a sampling of word problems whose solutions require the use of the four basic arithmetic operations: addition, subtraction, multiplication and division and use this opportunity to further hone our skill at modelling word problems as algebraic equations, analyzing and synthesizing these equations, and relating algebraic solutions to step-by-step solutions.

2.1 Addition, Subtraction and Multiplication

In this section we take a closer look at problems whose solutions require the use of addition, subtraction and multiplication. We begin with a simple problem.

Problem

You bought 3 pairs of socks at $4.20/pair and 2 hats at $7.25/hat. What amount of money did you spend in total?

For a step-by-step solution, we begin by working out a set of steps that would solve this problem. One possibility is to calculate the amount of money spent on the socks and the amount of money spent on the hats, and proceed from there to calculate total amount of money spent. A step-by-step solution can now be generated by listing the steps above with annotations and calculations.[1,2]

A Step-by-Step Solution

1. Calculate amount of money spent on the socks.

$$4.20 \times 3 \ = \ 12.60$$

I spent \$12.60 on the socks.

2. Calculate amount of money spent on the hats.

$$7.25 \times 2 \ = \ 14.50$$

I spent \$14.50 on the hats.

3. Calculate total amount of money spent.

$$12.60 \ + \ 14.50 \ = \ 27.10$$

In total I spent \$27.10.

For an algebraic solution, we begin by relating the values of the various quantities involved[3] in the simplest possible manner. One simple relationship would state that *total amount of money spent is equal to the sum of the amount of money spent on the socks and amount of money spent on the hats.* This leads to the following solution.

---

[1]The steps in the step-by-step solutions given in this textbook correspond to the main operation in the expression on the right side of the equation in the particular model that has the symbol for the quantity whose value is sought isolated on its left side. The initial set of steps is devoted to the evaluation of the arguments of the main operation and the last step corresponds to the execution of the main operation.

Substeps correspond to the next main operation within each step. As before, the initial set of steps is devoted to the evaluation of the arguments of the main operation and the last step corresponds to the execution of the main operation.

[2]Numerical values that represent amounts of money must include two decimal digits unless both decimal digits are 0 in which case one may exclude the decimal digits altogether. This means that we write \$12, \$12.00 (preferred) and \$12.10 but not \$12.1. In addition, at times, when working with problems that involve money, one arrives at a result that contains more than two decimal digits. Unless the requirements in your field of activity state otherwise, such numerical values should be rounded to two decimal digits.

[3]Including the value of the quantity whose value is sought.

An Algebraic Solution

a total amount of money spent (\$)

$a = 4.20 \times 3 + 7.25 \times 2$

$a = 12.60 + 14.50$

$a = 27.10$

In total, I spent \$27.10.

Our next problem involves chain multiplication.

Problem

You order 20 boxes of printer paper. Each box contains 10 packages of printer paper at \$0.95/package.[4] What is the cost of the order?

To solve the problem using the step-by-step technique, we could start by calculating the total number of packages in all the boxes, and then calculate the cost of the order. This leads to the following step-by-step solution:

A Step-by-Step Solution

1. Calculate the total number of packages.

$10 \times 20 = 200$

There are 200 packages in total.

2. Calculate the cost of the order.

$0.95 \times 200 = 190$

The order costs \$190.

The algebraic technique displays the full logic behind the solution on one line. This is shown below.

---

[4]As a matter of style, we write 0.95 and not .95. This adheres to the good writing practice that, when a number is written using the decimal notation, if the whole part is 0, the 0 should be written down. This is also a requirement of the International System of Units, SI, which is the default measurement system used in the sciences. The explicit 0 removes any doubts and forces the decimal point to stand out. Note that failure to account for the presence of the decimal point in a number written in decimal notation results in a value that is at least 10 times larger.

An Algebraic Solution

c cost of the order (\$)
$c = 0.95 \times 10 \times 20$

$c = 190$

The order costs \$190.00.

The order of factors in the expression on the right side of the equality in the model is important. Chain multiplication can be decoded from right to left. To make sense of

$$0.95 \times 10 \times 20$$

we begin by interpreting the rightmost factor, i.e., 20. This is the number of boxes:

$$\underbrace{20}_{\text{number of boxes}}$$

Next, we move back one factor and interpret the quantity 10×20. As illustrated below, this is the number of packages:

$$\underbrace{10 \quad \times \quad 20}_{\text{number of packages}}$$

We move back one more factor to get $0.95 \times 10 \times 20$. As shown below, this represents the cost of the order:

cost of a package	times	number of packages		
0.95	$\times$	10	$\times$	20

$$\underbrace{0.95 \quad \times \quad 10 \quad \times \quad 20}_{\text{cost of the order}}$$

Problem

Calculate the mass of 1 molecule of H_2SO_4 in amu given that the atomic mass of H is 1.008 amu/atom, the mass of an atom of S is 32.07 amu and the atomic mass of O is 16.00 amu/atom.

The notation H_2SO_4 tells us that a molecule of H_2SO_4 contains 2 atoms of H, 1 atom of S, and 4 atoms of O. Therefore, to find the mass of a molecule of H_2SO_4 we need to add the masses of 2 atoms of H, 1 atom of S, and 4 atoms of O. A step-by-step solution follows.

A Step-by-Step Solution

1. Calculate the mass of the H atoms.

   $$1.008 \times 2 \; = \; 2.016$$

   The H atoms have a mass of 2.016 amu.

2. Calculate the mass of the O atoms.

   $$16.00 \times 4 \; = \; 64$$

   The O atoms have a mass of 64 amu.

3. Calculate the mass of a molecule of H_2SO_4.

   $$2.016 \; + \; 32.07 \; + \; 64 \; = \; 98.086$$

   A molecule of H_2SO_4 has a mass of 98.086 amu.

For the algebraic solution, we begin by relating the various quantities involved in the simplest possible manner. One such relationship states that the mass of a molecule of H_2SO_4 is equal to the sum of the masses of the H, S and O atoms in it. An algebraic solution follows.

An Algebraic Solution

$$m \quad \text{mass of a molecule of } H_2SO_4 \text{ (amu)}$$
$$m \; = \; 1.008 \times 2 \; + \; 32.07 \; + \; 16.00 \times 4$$

$$m \; = \; 2.016 \; + \; 32.07 \; + \; 64$$
$$m \; = \; 98.086$$

The mass of a molecule of H_2SO_4 is 98.086 amu.

As a matter of style, we order our terms so that they evaluate the masses of the atoms in the order in which they appear in the chemical formula H_2SO_4.

Problem

Calculate the mass of 1 mol of H_2SO_4 in grams given that the molar mass of H is 1.008 g/mol, the mass of 1 mol of S is 32.07 g and the molar mass of O is 16.00 g/mol.

Since 1 mol of H_2SO_4 generates 2 mol of H atoms, 1 mol of S atoms, and 4 mol of O atoms,[5] to find the mass of 1 mol of H_2SO_4 we need to add the masses of 2 mol of H atoms, 1 mol of S atoms, and 4 mol of O atoms. A step-by-step solution follows.

A Step-by-Step Solution

1. Calculate the mass of the H atoms.

$$1.008 \times 2 = 2.016$$

   The H atoms have a mass of 2.016 g.

2. Calculate the mass of the O atoms.

$$16.00 \times 4 = 64$$

   The O atoms have a mass of 64 g.

3. Calculate the mass of 1 mol of H_2SO_4.

$$2.016 + 32.07 + 64.00 = 98.086$$

   The mass of 1 mol of H_2SO_4 is 98.086 g.

For an algebraic solution, we begin by relating the various quantities involved in the simplest possible manner. One such relationship states that the mass of 1 mol of H_2SO_4 is equal to the sum of the masses of 2 mol of H atoms, 1 mol of S atoms, and 4 mol of O atoms. An algebraic solution follows.

An Algebraic Solution

m mass of 1 mol of H_2SO_4 (g)
$m = 1.008 \times 2 + 32.07 + 16.00 \times 4$

$m = 2.016 + 32.07 + 64$
$m = 98.086$

The mass of 1 mol of H_2SO_4 is 98.086 g.

The difference between the solutions given for the preceding two problems is in the sets of units used. If we were to include units in our equations, the

---

[5] See Appendix B for a detailed discussion. Briefly, since each molecule of H_2SO_4, if broken up into atoms, generates twice as many H atoms, as many S atoms, and four times as many O atoms, 1 mol of H_2SO_4, if broken up into atoms, generates 2 mol of H atoms, 1 mol of S atoms, and 4 mol of O atoms.

first term on the right side of the equation in the model for the first problem would read as

    1.008 amu/atom × 2 atoms

which evaluates to 2.016 amu while the first term on the right side of the equation in the model for the second problem would read as

    1.008 g/mol × 2 mol

which evaluates to 2.016 g.

    We will present one more problem.

Problem

On average we can extract 4 Cal of energy from 1 g of carbohydrate, 9 Cal of energy from 1 g of fat and 4 Cal of energy from 1 g of protein. A chocolate bar contains 78.6 grams of carbohydrate, 16.7 grams of fat, and 2.0 grams of protein. How much energy can be extracted from the chocolate bar?

We can solve this problem using the step-by-step approach by calculating the energy content of carbohydrates in the chocolate bar, the energy content of fat in the chocolate bar, the energy content of proteins in the chocolate bar and then add these together to find the total energy content of the chocolate bar. This line of reasoning leads to the following step-by-step solution.

A Step-by-Step Solution

1. Calculate the energy contribution from carbohydrates.

    $4 \times 78.6 = 314.4$

The carbohydrates contribute 314.4 Cal to the energy content of the chocolate bar.

2. Calculate the energy contribution from fat.

    $9 \times 16.7 = 150.3$

The fat contributes 150.3 Cal to the energy content of the chocolate bar.

3. Calculate the energy contribution from proteins.

$$4 \times 2 = 8$$

The proteins contribute 8 Cal to the energy content of the chocolate bar.

4. Calculate the energy content of the chocolate bar.

$$314.4 + 150.3 + 8 = 472.7$$

The chocolate bar contains 472.7 Cal of energy.

For an algebraic solution, we relate the values of the various quantities involved in the word problem in the simplest possible manner. One such relationship states that the energy content of the chocolate bar is equal to the sum of the energy contributions from its carbohydrate, fat and protein content. This leads to the following algebraic solution.

An Algebraic Solution

E energy content of the chocolate bar (Cal)
$E = 4 \times 78.6 + 9 \times 16.7 + 4 \times 2$

$E = 314.4 + 150.3 + 8$
$E = 472.7$

The chocolate bar contains 472.7 Cal of energy.

Exercise Set 2.1

1. Solve each of the following problems using the step-by-step and algebraic techniques.

   a. You bought 6 pens at $2.50/pen and 8 notebooks at $4.99/notebook. What amount of money did you spend?

   b. You bought 6.5 metres of fabric at $1.20/m and 3 rolls of string at $2.00/roll. What amount of money did you spend?

   c. You ordered 20 boxes of printer paper. Each box contains 18 packages of paper at $2.99/package. What was the cost of the order?

   d. You ordered 20 boxes of surgical gloves at $5.20/box and 32 boxes of masks at $10.99/box. How much did the order cost?

e. You start with 400 mL of a medication. You use medication at a rate of 5.65 mL/min for 12 min. What volume of the medication do you have left?

f. You start with 180 L of a salt solution. You use salt solution at a rate of 2.4 L/h for 3 h. What volume of the salt solution do you have left?

g. Calculate the mass of a molecule of C_2H_6 in amu. The atomic masses of C and H are 12.01 amu/atom and 1.008 amu/atom respectively.

h. Calculate the mass of 1 mole of C_2H_6 in grams. The molar masses of C and H are 12.01 g/mol and 1.008 g/mol respectively.

i. Calculate the mass of 1 mole of SO_2 in grams. The mass of 1 mol of S is 32.07 g. The molar mass of O is 16.00 g/mol.

j. Calculate the mass of a molecule of CH_4 in amu. The mass of an atom of C is 12.01 amu. The atomic mass of H is 1.008 amu/atom.

k. Calculate the mass of a molecule of H_2O in amu. The atomic mass of H is 1.008 amu/atom. An atom of O has a mass of 16.00 amu.

l. Calculate the mass of 1 mol of CO_2 in grams. The mass of 1 mol of C is 12.01 g. The molar mass of O is 16.00 g/mol.

m. Calculate the mass of 1 mol of ozone, O_3, in grams. The molar mass of O is 16.00 g/mol.

n. On average we can extract 4 Cal/g of carbohydrate, 9 Cal/g of fat and 4 Cal/g of protein. A glass of milk contains 12.5 g of carbohydrate, 1.4 g of fat, and 4.2 g of protein. What is its energy content in Cal?

o. On average we can extract 4 Cal/g of carbohydrate and 9 Cal/g of fat. A cup of yogurt contains 16 grams of carbohydrate and 4.2 grams of fat. Calculate the energy content of the cup of yogurt in Calories.

p. For breakfast Marco had 3 slices of buttered bread at 37 Cal/slice, 2 spoons of jam at 52 Cal/spoon, and a cup of tea containing 20 Cal. He rested for 2 hours, burning on average 80 Cal/h. Then he exercised for 0.5 of an hour, burning Calories at a rate of 320 Cal/h. What is Marco's net intake of energy since breakfast?

q. A breakfast deal contains 3 slices of cheese at 80 Cal/slice, 2 slices of toast at 40 Cal/slice, 2 cups of coffee at 90 Cal/cup and scrambled eggs containing 120 Cal. Calculate the energy content of the breakfast deal.

2. In each case write a word problem whose solution can be modelled as the given equation. Replace the abstract quantity symbol, x, on the left side of the equation with the relevant concrete quantity symbol and write the model for the word problem.

a. $x = 4 \times 2 + 15 \times 3$
b. $x = 12.99 + 2.5 \times 4$
c. $x = 1020 - 430 \times 2 + 500$
d. $x = 12.43 \times 3.1 + 9.2 \times 4.2 + 3.8 \times 2.5$
e. $x = 20.5 - 2.5 + 4.2$
f. $x = 3 \times 4 \times 5$
g. $x = 5.99 \times 3.1 \times 2$

2.2   Multiplication and Brackets

Although in basic mathematics the notation $\times$ is commonly used to represent multiplication with, in algebra the use of brackets for this purpose is generally preferred. In this context the term **brackets** refers to either **parentheses**, i.e., ( ), or **square brackets**, i.e., [ ].

We can use brackets to write the product

$$4 \times 7$$

as

$$(4)\,(7)$$

Although the use of brackets for the representation of multiplication in algebraic expressions facilitates their analysis, their use in expressions that involve nested multiplication[6] quickly leads to a deterioration of the readability of the expression. For this reason, we adopt conventions that tend to minimize the use of brackets.[7] These conventions are listed below.

The first convention aims to reduce the use of brackets by dropping the brackets around the first factor in expressions that involve chain multiplication. We, therefore, write $4\,(7)$ in place of $(4)\,(7)$ and we write $-3\,(-2)\,(5)$ in place of $(-3)\,(-2)\,(5)$.

The second convention aims to reduce the number of brackets by replacing the brackets at the lowest level with the $\times$ symbol[8] in cases when all factors

---

[6]By *nested multiplication* we mean multiplication within multiplication; a scenario that requires the use of brackets within brackets.

[7]Unless semantics require that we break convention, these conventions will be applied in the writing of expressions and equations that appear in this book.

[8]We do not recommend the use of the **multiplication dot** for the representation of multiplication in mathematical expressions as its use between numerical values may lead to interpretation errors. While it is true that the multiplication dot is centered along the font height while the decimal point sits on the base line (as in $4 \cdot 5$ which implies 4×5 vs. 4.5 which implies *four and five tenths*), its use still requires that the reader be aware of the possibility of misinterpretation especially in expressions that are handwritten.

except perhaps the first factor are positive. We, therefore, write $2 \times 8 \times 4$ in place of $2\,(8)\,(4)$ and we write

$$3\,(2 \times 4 \;+\; 3 \times 2 \times 5)$$

in place of

$$3\Big[2\,(4) \;+\; 3\,(2)\,(5)\Big]$$

As a matter of style we will allow the use of either brackets or the $\times$ symbol (preferred) at the lowest level but not a mix of the two. We, therefore, write $4\,(7)\,(6)$ and $4 \times 7 \times 6$ (preferred), but not $4 \times 7\,(6)$ or $4\,(7) \times 6$. The latter two presentations suffer from the point of view of readability and may leave the wrong impression that the chain multiplication is split.

Note also that using brackets to represent multiplication with does not imply that the brackets enclose the full expression, e.g., we do *not* write (4×7) but $4\,(7)$.[9] The notation $4\,(7)$ is acceptable as the brackets are being multiplied on the outside by 4. However, there are no values multiplying the expression (4×7) from outside and, therefore, the brackets are not needed.

With these conventions in mind, let us now present a few problems whose algebraic solutions require the use of brackets.[10]

Problem

Calculate the mass of 20 molecules of $C_6H_{12}O_6$ in amu. The atomic masses of C, H and O are 12.01 amu/atom, 1.008 amu/atom and 16.00 amu/atom respectively.

We can solve this problem by calculating the mass of each type of atom in a molecule of $C_6H_{12}O_6$, use this information to calculate the mass of one molecule of $C_6H_{12}O_6$, and use the mass of one molecule of $C_6H_{12}O_6$ to calculate the mass of all molecules of $C_6H_{12}O_6$. A step-by-step solution follows.

---

[9]To prevent overuse of brackets in this manner, check to see whether any brackets used in an expression are multiplied on the outside by a value. If so, the use of brackets is appropriate otherwise it is not.

An exception to this rule appears in the modelling of problems whose solution requires that we subtract a sum from another value. In such cases the sum is placed inside brackets with no value multiplying the brackets from the outside. See the problem at the end of this chapter for an example.

It should also be noted that, later in the text, brackets will be further burdened with the task of grouping the arguments of higher order operations.

[10]The reader should be aware that it is not uncommon to come across expressions written by individuals who do not observe style as we do. It is, therefore, important that the reader should train herself or himself to paraphrase expressions that are written by others into ones that adhere to the conventions that we have listed above. This restores clarity and aids analysis.

A Step-by-Step Solution

1. Calculate the mass of a molecule of $C_6H_{12}O_6$.

   a. Calculate the mass of the C atoms in a molecule of $C_6H_{12}O_6$.

   $$12.01 \times 6 \ = \ 72.06$$

   The C atoms have a mass of 72.06 amu.

   b. Calculate the mass of the H atoms in a molecule of $C_6H_{12}O_6$.

   $$1.008 \times 12 \ = \ 12.096$$

   The H atoms have a mass of 12.096 amu.

   c. Calculate the mass of the O atoms in a molecule of $C_6H_{12}O_6$.

   $$16.00 \times 6 \ = \ 96$$

   The O atoms have a mass of 96 amu.

   d. Calculate the mass of a molecule of $C_6H_{12}O_6$.

   $$72.06 \ + \ 12.096 \ + \ 96 \ = \ 180.156$$

   A molecule of $C_6H_{12}O_6$ has a mass of 180.156 amu.

2. Calculate the mass of $C_6H_{12}O_6$ molecules.

   $$180.156 \times 20 \ = \ 3603.12$$

   The $C_6H_{12}O_6$ molecules have a mass of 3603.12 amu.

For the algebraic technique, we start by relating the values of the various quantities involved in the simplest possible manner. One such relationship states that the mass of the molecules of $C_6H_{12}O_6$ is equal to the product of the number of $C_6H_{12}O_6$ molecules and the mass of one molecule of $C_6H_{12}O_6$. This leads to the following algebraic solution.[11]

---

[11]One might wonder why we did not follow our earlier convention and write

$$\big(6 \times 12.01 \ + \ 12 \times 1.008 \ + \ 6 \times 16\big)(20)$$

where the rate is listed first. The reason is that the presentation that we have adopted eliminates the need for placing brackets around 20. We adopt the convention that if a numerical value multiplies brackets, the numerical value is written on the left side of the brackets. This convention has precedence over our earlier conventions; however, feel free to break any conventions if semantic requirements dictate that you should.

An Algebraic Solution

m mass of the molecules of $C_6H_{12}O_6$ (amu)
$m = 20(12.01 \times 6 + 1.008 \times 12 + 16.00 \times 6)$

$m = 20(72.06 + 12.096 + 96)$
$m = 20 \times 180.156$
$m = 3603.12$

The $C_6H_{12}O_6$ molecules have a mass of 3603.12 amu.

The expression

$20(12.01 \times 6 + 1.008 \times 12 + 16.00 \times 6)$

on the right side of the equation in the algebraic model has only one term. The single term breaks into two factors. These are

20

and

$12.01 \times 6 + 1.008 \times 12 + 16.00 \times 6$

Factors that contain more than one term, such as the one above, are enclosed within brackets.[12]

Note also that we have adhered to our conventions on minimizing the use of brackets. Without these conventions the expression on the right side of the equality in the model would read as

$$\left[12.01(12.01) + 1.008(12) + 16.00(6)\right](20)$$

We are already running into nested brackets and this will get worse with the introduction of another layer of brackets, a common occurrence in practical applications – See below.

To evaluate the expression on the right side of the equation

$m = 20(12.01 \times 6 + 1.008 \times 12 + 16.00 \times 6)$

we must evaluate the factors and then multiply them together. The first factor is simply 20 and cannot be simplified further. The second factor, the expression inside brackets, breaks into three terms. These are 12.01×6, 1.008×12 and 16.00×6. We evaluate these terms in one step to get

$m = 20(72.06 + 12.096 + 96)$

---

[12]If brackets are used according to the guidelines that we have listed, then each set of brackets constitutes a factor within the term in which it resides.

Next, we add the terms within the bracketed factor to get

$$m = 20 \times 180.156$$

The rest follows.

Problem

Calculate the mass of a molecule of $(NH_4)_3\,CO_2$ in amu. The atomic masses of N, H, and O are 14.01 amu/atom, 1.008 amu/atom, and 16.00 amu/atom respectively. The mass of an atom of C is 12.01 amu.

The notation $(NH_4)_3\,CO_2$ implies that the molecule contains 3 groups of NH_4, an atom of C and 2 atoms of O. To find the mass of a molecule of $(NH_4)_3CO_2$ we can add the masses of 3 groups of NH_4, an atom of C and 2 atoms of O. Here is a step-by-step solution.

A Step-by-Step Solution

1. Calculate the mass of NH_4 groups in a molecule of $(NH_4)_3\,CO_2$.

    a. Calculate the mass of one group of NH_4.

        i. Calculate the mass of H atoms in a group of NH_4.

            $$1.008 \times 4 = 4.032$$

            The H atoms have a mass of 4.032 amu.

        ii. Calculate the mass of one group of NH_4.

            $$14.01 + 4.032 = 18.042$$

            An NH_4 group has a mass of 18.042 amu.

    b. Calculate the mass of NH_4 groups.

        $$18.042 \times 3 = 54.126$$

        The NH_4 groups have a mass of 54.126 amu.

2. Calculate the mass of the O atoms in a molecule of $(NH_4)_3\,CO_2$.

        $$16.00 \times 2 = 32$$

        The O atoms have a mass of 32 amu.

3. Calculate the mass of a molecule of $(NH_4)_3\,CO_2$.

$$54.126 + 12.01 + 32 = 98.136$$

A molecule of $(NH_4)_3\,CO_2$ has a mass of 98.136 amu.

For the algebraic presentation we begin by relating the various quantities involved in the simplest possible manner. One such relationship states that the mass of a molecule of $(NH_4)_3CO_2$ is equal to the sum of the masses of the groups of NH_4, the atom of C, and the atoms of O. This leads to the following algebraic solution.

An Algebraic Solution

m mass of a molecule of $(NH_4)_3\,CO_2$ (amu)
$m = 3\,(14.01 + 1.008 \times 4) + 12.01 + 16.00 \times 2$

$m = 3\,(14.01 + 4.032) + 12.01 + 32$
$m = 3 \times 18.042 + 12.01 + 32$
$m = 54.126 + 12.01 + 32$
$m = 98.136$

The mass of a molecule of $(NH_4)_3\,CO_2$ is 98.136 amu.

The expression on the right side of the equation that models the problem above contains three terms. These are boxed below.

$$\boxed{3(14.01 + 1.008 \times 4)} + \boxed{12.01} + \boxed{16.00 \times 2}$$

The first term represents the mass of the groups of NH_4 in a molecule of $(NH_4)_3CO_2$; the second term represents the mass of an atom of C in a molecule of $(NH_4)_3CO_2$; and the last term represents the mass of the O atoms in a molecule of $(NH_4)_3CO_2$.

We start to work on all the terms simultaneously. The first term contains two factors: 3, which is already simplified, and the expression inside brackets. This expression breaks into two terms. We begin by evaluating these terms.

$$\boxed{3\,(14.01 + 4.032)} + \cdots$$

Next we have the term 12.01 which is already as simplified as possible. We write

$$\boxed{3\,(14.01 + 4.032)} + \boxed{12.01} + \cdots$$

The third term analyzes into two factors: 16.00 and 2. Since both factors have been simplified as much as possible, we proceed to multiply them to get

$$\boxed{3(14.01 \,+\, 4.032)} \,+\, \boxed{12.01} \,+\, \boxed{32}$$

The first term still needs simplification. We add the mass of the N atom and the mass of the H atoms to get

$$\boxed{3 \times 18.032} \,+\, \boxed{12.01} \,+\, \boxed{32}$$

This simplifies to

$$54.096 \,+\, 12.01 \,+\, 32$$

The rest is simple.

Problem

Calculate the mass of 20 molecules of $(NH_4)_3\,CO_2$ in amu. The atomic masses of N, H, and O are 14.01 amu/atom, 1.008 amu/atom, and 16.00 amu/atom respectively. The mass of an atom of C is 12.01 amu.

We will present an algebraic solution for this problem. The algebraic solution calculates the mass of the $(NH_4)_3\,CO_2$ molecules by calculating the mass of one molecule of $(NH_4)_3\,CO_2$ as in the previous problem and multiplying the result by the number of $(NH_4)_3\,CO_2$ molecules.

An Algebraic Solution

m mass of $(NH_4)_3\,CO_2$ molecules (amu)
$$m \,=\, 20\Big[3(14.01 \,+\, 1.008 \times 4) \,+\, 12.01 \,+\, 16.00 \times 2\Big]$$

$$m \,=\, 20\Big[3(14.01 \,+\, 4.032) \,+\, 12.01 \,+\, 32\Big]$$
$$m \,=\, 20(3 \times 18.042 \,+\, 12.01 \,+\, 32)$$
$$m \,=\, 20(54.126 \,+\, 12.01 \,+\, 32)$$
$$m \,=\, 20 \times 98.136$$
$$m \,=\, 1962.72$$

The $(NH_4)_3\,CO_2$ molecules have a mass of 1962.72 amu.

The expression inside square brackets represents the mass of a molecule of $(NH_4)_3CO_2$ (See the previous problem). This is placed inside brackets and multiplied by 20 to generate the mass of 20 molecules of $(NH_4)_3CO_2$.

When brackets are nested, readability can be improved by alternating between parentheses, i.e., ( ), and square brackets, i.e., [ ], with innermost brackets represented as parentheses by convention. We can improve readability further by using larger bracket sizes for outer brackets. The combination of alternating bracket shapes and sizes provides a powerful visual cue, allowing the practitioner to pair related open/close brackets at a glance.

We end this section with one more problem.

Problem

You had \$126.50. You bought 4 notebooks at \$7.20/notebook, a binder for \$12.59 and 2 pens at \$0.90/pen. How much do you have left?

One way to solve this problem is to begin with the amount at the beginning and then successively subtract the cost of the notebooks, the cost of the binder and the cost of the pens. This would result in the following model.

$$a = 126.50 - 7.20 \times 4 - 12.59 - 0.90 \times 2$$

where a represents the value of the amount of money left. However, most of us have been taught at an early age to solve such problems by first adding the amounts of money paid for the notebooks, the binder and the pens, and then subtract this total sum from amount of money that we started with. Such a model would require the use of brackets as shown below.

An Algebraic Solution

a amount I have left after purchases (\$)
$a = 126.50 - (7.20 \times 4 + 12.59 + 0.90 \times 2)$

$a = 126.50 - (28.80 + 12.59 + 1.80)$
$a = 126.50 - 43.19$
$a = 83.31$

I have \$83.31 left after purchases.

The advantage of this model over the model that uses successive subtractions is that it employs fewer subtractions, in fact, only a single subtraction. There is a definite advantage to this approach when calculations are performed by hand.

One might wonder whether the model that uses successive subtractions might be better on the account that it uses fewer brackets and that the brackets in the model above are not multiplied by a value from the outside. It is important for the reader to note that, from time to time, in practice the conventions that we have set up are broken, but always for good reason. In the present case, as an example, semantics of the solution require the use of brackets and this need overrides our convention on minimizing the use of brackets. Math is, first and foremost, a reasoning activity and therefore it is the needs of logic that dictate whether brackets should be used not vice versa. If we were to give priority to syntactic preferences over semantic needs, then algebra would hinder, rather than facilitate, our reasoning.

Exercise Set 2.2

Solve each of the following problems using the step-by-step and algebraic techniques.

1. You had $58. You bought a calculator for $16.99 and 3 pens at $2.00/pen. How much do you have left?

2. You had $420 in your bank. You made 3 payments of $52.50 each and 2 payments of $30.99 each. How much do you have left?

3. You start with 657 mL of a medication. You give 20.5 mL of the medication to each of 8 patients, 12.5 ml of medication to each of 12 patients, and 3.5 mL of the medication to each of 18 patients. How much medication do you have left?

4. Maria started with 8.1 L of a salt solution. This morning she used salt solution at a rate of 1.25 L/h for 1.5 h. This afternoon she used salt solution at a rate of 0.57 L/h for 2.5 h. What volume of the salt solution does Maria have left?

5. Calculate the mass of 42 molecules of SO_2 in amu. The atomic masses of S and O are 32.07 amu/atom and 16.00 amu/atom respectively.

6. Calculate the mass of 42 moles of SO_2 in grams. The molar masses of S and O are 32.07 g/mol and 16.00 g/mol respectively.

7. Calculate the mass of a molecule of $Ca(OH)_2$ in amu. The mass of an atom Ca, an atom of O and an atom of H is 40.08 amu, 16.00 amu and 1.008 amu.

8. Calculate the mass of 1 mol of $Li_2(CO_3)_4$ in grams. The molar masses of Li, C and O are 6.941 g/mol, 12.01 g/mol and 16.00 g/mol respectively.

9. Calculate the mass of 4.8 moles of $Na_4(PO_4)_3$ in grams. The molar masses of Na, P and O are 22.99 g/mol, 30.97 g/mol and 16.00 g/mol respectively.

10. Calculate the mass of 158 molecules of $(NH_4)_3\,PO_4$ in amu. The mass of an atom of N, an atom of H, and atom of P and an atom of O are 14.01 amu, 1.008 amu, 30.97 amu and 16.00 amu respectively.

2.3   Division and the Horizontal Line

We have noted that the analysis of an algebraic expression proceeds along the lines of additions and subtractions (which set the terms) and next, within each term, multiplications (which set the factors). In this view, higher order operations (division, exponents, roots, etc.) are found within factors.

To represent division itself, we use the horizontal line. The horizontal line has many advantages over other notations such as $\div$ or / for the representation of division. One such advantage[13] is that the horizontal line naturally groups its divisor while use of alternative notations often requires the use of brackets for this purpose.[14] As an example, using the horizontal line to denote division with, we can use the expression

$$\frac{14}{2+5}$$

to communicate the need to compute the quotient of 14 and the sum of 2 and 5 while the use of either $\div$ or the / would require the use of brackets:

$$14 \div (2 + 5)$$

and

$$14/(2 + 5)$$

Removal of the brackets changes the expression. Note that

$$14 \div (2 + 5) = 14 \div 7$$
$$= 2$$

while

$$14 \div 2 + 5 = 7 + 5$$
$$= 12$$

---

[13]For more on the advantages of the horizontal line over other notations for division please see the companion textbook *Style in Technical Math* by the author.

[14]This conflicts with our earlier insistence that brackets should only be used to represent multiplication with.

In this textbook we will use the horizontal line as the symbol to represent division with.[15]

Having discussed notation, let us now present a few problems whose solutions require the use of division.

Problem

You made \$120 on Monday, \$140 on Tuesday, \$80 on Wednesday and \$150 on Thursday. On average, what amount of money did you make per day over the four day period?

To solve this problem in a step-by-step manner, we calculate the total amount of money made over the four-day period and divide this by the number of days.

A Step-by-Step Solution

1. Find total amount of money made.

$$120 + 140 + 80 + 150 = 490$$

In total you made \$490.

2. Calculate average amount of money made per day.

$$490 \div 4 = 122.50$$

On average you made \$122.50/day over the four day period.

For an algebraic solution, we begin by relating the values of the various quantities involved in the simplest possible manner. One such relationship states that, to calculate the average amount of money made per day, we can calculate the total amount of money made over the four-day period and divide this by the number of days. This leads to the following algebraic solution.[16]

---

[15]The reader should note that others may not observe style as we do. Therefore, it is important that the reader should train herself or himself to paraphrase expressions that are written by others and convert $\div$ and / signs to the horizontal line. This brings clarity and aids analysis.

[16]In this solution the line over the quantity symbol a represents an average value.

An Algebraic Solution

$\bar{a}$ average amount of money made per day (\$/day)

$$\bar{a} = \frac{120 + 140 + 80 + 150}{4}$$

$$\bar{a} = \frac{490}{4}$$

$$\bar{a} = 122.50$$

On average you made \$122.50/day over the four day period.

In some average problems one or more pieces of data to be averaged repeat. Here is an example.

Problem

You have 3 Type A packages, 4 Type B packages and 2 Type C packages. The masses of a Type A package, a Type B package and a Type C package are 14.2 g, 26.8 g and 50.0 g respectively. What is the average mass of a package?

We can solve this problem by calculating the mass of Type A, Type B and Type C packages, add these to find the total mass of all the packages and then divide the result by the number of packages. A step-by-step solution follows.

A Step-by-Step Solution

1. Find the mass of all packages.

   a. Calculate the mass of Type A packages.

   $$14.2 \times 3 = 42.6$$

   Type A packages have a combined mass of 42.6 g.

   b. Calculate the mass of Type B packages.

   $$26.8 \times 4 = 107.2$$

   Type B packages have a combined mass of 107.2 g.

   c. Calculate the mass of Type C packages.

   $$50.0 \times 2 = 100$$

   Type C packages have a combined mass of 100 g.

  d. Calculate the mass of all the packages.

$$42.6 + 107.2 + 100 = 249.8$$

  The combined mass of all the packages is 249.8 g.

2. Calculate the total number of packages.

$$3 + 4 + 2 = 9$$

There are 9 packages in total.

3. Calculate the average mass of a package.

$$249.8 \div 9 = 27.7555\ldots$$

The average mass of a package is 27.8 g/package.

For the algebraic technique, we reason by noting that the average mass of a package can be calculated by dividing the total mass by the number of packages.

An Algebraic Solution

$\overline{m}$ average mass of a package (g/package)

$$\overline{m} = \frac{14.2 \times 3 + 26.8 \times 4 + 50.0 \times 2}{3 + 4 + 2}$$

$$\overline{m} = \frac{42.6 + 107.2 + 100}{9}$$

$$\overline{m} = \frac{249.8}{9}$$

$$\overline{m} = 27.7555\ldots$$

The average mass of a package is 27.8 g/package.

A few comments are in order.

The dividend in the expression on the right side of the equation that models the problem in the algebraic solution uses multiplication to speed up the process of calculating the total mass of the packages, i.e., we write

$$14.2 \times 3 + 26.8 \times 4 + 50.0 \times 2$$

in place of the longer

$$14.2 + 14.2 + 14.2 + 26.8 + 26.8 + 26.8 + 26.8 + 50.0 + 50.0$$

The former formulation is better as it is shorter and more efficient.[17]

The divisor in the expression on the right side of the equation that models the problem in the algebraic solution counts the total number of packages. There are 3 Type A packages, 4 Type B packages and 2 Type C packages for a total of $3 + 4 + 2$ packages.

In an average problem the number of times a piece of data is repeated is referred to as that piece of data's *weight*. As an example, since the mass of a Type A package (i.e., 14.2 g) is repeated 3 times in the calculation of the average mass in the problem above, we refer to 3 as the *weight* of 14.2. And since the mass of a Type B package (i.e., 26.8 g) is repeated 4 times in the calculation of the average mass in the problem above, we refer to 4 as the weight of 26.8. In general, if a piece of data is counted n times in the calculation of an average value, we refer to n as the weight of that piece of data. The significance of the weights is in their effect on the average value: The higher the weight of a piece of data, the stronger the pull on the average value towards that piece of data.

Averages whose calculations involve weights are called **weighted averages** with the **weights** representing the number of times a piece of data is repeated in the calculation of the average value. These weights also appear in the divisor where we count the total number of data. This is illustrated below where we have used bold font to represent the weights:

$$\overline{m} = \frac{14.2 \times \mathbf{3} + 26.8 \times \mathbf{4} + 50.0 \times \mathbf{2}}{\mathbf{3} + \mathbf{4} + \mathbf{2}}$$

As you will see from the examples that follow, this pattern is present in the solution of all weighted average problems.

In addition to the above, it should be noted that an average value of a set of data cannot be less than the minimum value in the set of data or greater than the maximum value in the set of data.

Here is a problem from physics.

Problem

This morning you drove at a speed of 120 km/h for 2.5 h. This afternoon you drove at a speed of 90 km/h for 1.5 h. What was your average speed today?

Speed is, of course, defined as distance covered in unit time. As an example, a speed of 120 km/h implies that the object covers a distance of 120

---

[17]Imagine having 72 Type A packages each with a mass of 14.2 g. The former formulation would write the mass of these packages as 14.2×72 whereas the latter formulation would require that we add 14.2 by itself 72 times.

km in 1 h. To calculate average speed, then, we can calculate total distance covered and divide this by total travel time.

In addition to the above, the distance covered by an object moving at constant speed is given by the product of its speed and travel time. As an example, if an object moves at a speed of 120 km/h for 2.5 h, then it covers a distance of 120 × 2.5 or 300 km.

Here is a step-by-step solution.

A Step-by-Step Solution

1. Calculate total distance covered today.

   a. Calculate distance covered this morning.

   $$120 \times 2.5 \ = \ 300$$

   I covered a distance of 300 km this morning.

   b. Calculate distance covered this afternoon.

   $$90 \times 1.5 \ = \ 135$$

   I covered a distance of 135 km this afternoon.

   c. Calculate total distance covered today.

   $$300 \ + \ 135 \ = \ 435$$

   I covered a distance of 435 km today.

2. Calculate total travel time.

   $$2.5 \ + \ 1.5 \ = \ 4$$

   I drove for 4 h today.

3. Calculate my average speed today.

   $$435 \div 4 \ = \ 108.75$$

   My average speed today was 108.75 km/h.

Here is an algebraic solution.

An Algebraic Solution

$\overline{v}$ my average speed today (km/h)

$$\overline{v} = \frac{120 \times 2.5 + 90 \times 1.5}{2.5 + 1.5}$$

$$\overline{v} = \frac{300 + 135}{4}$$

$$\overline{v} = \frac{435}{4}$$

$$\overline{v} = 108.75$$

My average speed today was 108.75 km/h.

The unit of the quantity whose average value is sought can be used to determine what quantities should appear as the dividend and divisor of the expression on the right side of the equation that models an average problem. As an example, in the problem above we seek to find your average speed in km/h. The unit in the dividend of km/h, i.e., km, tells us that the quantity whose value is represented by the dividend of the expression on the right side of the equation that models the problem is one of distance, specifically, total distance covered. This is what $120 \times 2.5 + 90 \times 1.5$ computes. The first term, i.e., 120×2.5, computes distance covered this morning and the second term, i.e., 90×1.5, computes distance covered this afternoon. The unit in the divisor of km/h, i.e., h, tells us that the quantity whose value is represented by the divisor of the expression on the right side of the equation that models the problem is one of time, specifically, total travel time. This is what $2.5 + 1.5$ computes.

Note also that the pattern that we saw earlier in the model for the problem on the average mass of a package is also observed above in the model for the average speed problem. This is shown below with the weights displayed using bold font.[18]

$$\overline{v} = \frac{120 \times \mathbf{2.5} + 90 \times \mathbf{1.5}}{\mathbf{2.5} + \mathbf{1.5}}$$

Before moving on to our next problem, we refer the reader to Appendix C for background information on grading schemes.

The acronym GPA stands for *grade point average* and is an average based on grade points that a student earns after she or he completes one or more

---

[18]It is interesting that similar patterns arise in such different contexts; however, the magic fades somewhat when we note that in both cases we are calculating averages where some of the data repeat multiple times.

courses. This means that, if percent grades or letter grades are used in your report card, you need to convert them to grade points before you can calculate your GPA.

In addition to the scales commonly used to report course grades, we also need to discuss a concept called credit value. By definition, the *credit value of a semester-long course* is equal to the number of contact hours[19] for that course per week.[20]

Problem

Last term your friend took two semester-long courses: PHY101 and ENG101. PHY101 has a credit value of 5 and ENG101 has a credit value of 3. Your friend's final grades in PHY101 and ENG101 were a B, corresponding to a grade point value of 3.00, and a C, corresponding to a grade point value of 2.00, respectively. What was your friend's term GPA last term?

It is tempting to argue that to solve this problem, one should add the two grades and divide the result by 2. However, a little reflection shows that this is a not a fair average as it ignores the fact that students spend more time in the physics class (5 h/week) compared to the English class (3 h/week). If an hour of instruction in the physics class has the same learning value as an hour of instruction in the English class, then it makes sense that, to get a fair average, we should count the final grade in the physics course 5 times but count the final grade in the English course only 3 times.

Here is a step-by-step solution.

A Step-by-Step Solution

1. Calculate total grade points earned.

    a. Calculate grade points earned from the PHY101 course.

        $3.00 \times 5 = 15$

    My friend earned 15 grade points from the PHY101 course.

    b. Calculate grade points earned from the ENG101 course.

        $2.00 \times 3 = 6$

    My friend earned 6 grade points from the ENG101 course.

---

[19]Contact hours are hours of instruction where the professor and students physically meet as in the lecture room during lecture time.

[20]For a year-long course (i.e., one that is two semesters long), the credit value is set equal to twice the number of contact hours for that course per week.

c. Calculate total grade points earned.

$$15 + 6 = 21$$

In total, my friend earned 21 grade points.

2. Calculate total credit value.

$$5 + 3 = 8$$

The total credit value is 8.

3. Calculate my friend's term GPA last term.

$$21 \div 8 = 2.63$$

Last term, my friend had a term GPA of 2.63.

Here is an algebraic solution.

An Algebraic Solution

$$g \qquad \text{my friend's term GPA last term (1)}$$

$$g = \frac{3.00 \times 5 + 2.00 \times 3}{5 + 3}$$

$$g = \frac{15 + 6}{8}$$

$$g = \frac{21}{8}$$

$$g = 2.63$$

Last term, my friend had a term GPA of 2.63.

Note that if we had added the grades and divided the sum by 2, we would have arrived at the erroneous term GPA of 2.5. Relatively speaking, the difference between this GPA and the actual GPA is not small.[21]

As before, we refer to 5 as the weight of 3.00 (meaning that 3.00 is repeated 5 times in the calculation of the average) and we refer to 3 as the weight of 2.00 (meaning that 2.00 is repeated 3 times in the calculation of the average).

---

[21]The difference is off by roughly 5% compared to the correct value. Note that the difference can be considerably larger. Had the grades in the two courses been an A and a D respectively, the correct GPA would have been 2.88 while the value that results from the division of the sum of the grades by 2 would have been 2.5. This difference is off by roughly 13% of the correct value.

We remind the reader that the formula for the calculation of the GPA above is based on the assumption that an hour of instruction in any course has the same learning value as an hour of instruction in any other course. If in some application this assumption does not hold, then the formula given above for the calculation of the GPA would have to be modified to represent the true average.

It is our logic that tells us what the GPA formula should look like not vice versa.

Our next example relates to the definition of atomic mass.[22]

Problem

A sample of chlorine atoms from nature contains 76% ^{35}Cl and 24% ^{37}Cl.[23] Calculate the atomic mass of chlorine.

Atomic mass is defined as the average mass of the isotopes of an atom in a sample of that atom from nature. To calculate this average we can find the mass of all the atoms in the sample and divide this by the total number of atoms in the sample.

For the problem above, the sample contains 76% ^{35}Cl and 24% ^{37}Cl. We may interpret the rates as the number of atoms in a sample of 100 atoms. Therefore, we can argue that out of every 100 atoms in the sample, 76 are of type ^{35}Cl and 24 are of type ^{37}Cl. Since each ^{35}Cl atom has a mass of 35 amu, the mass of 76 ^{35}Cl atoms is given by the expression 35×76. Furthermore, since each ^{37}Cl atom has a mass of 37 amu, the mass of 24 ^{37}Cl atoms is given by the expression 37×24. The total mass of 100 atoms from the sample is, therefore, given by the expression $35 \times 76 + 37 \times 24$.

The total number of atoms is, of course, $76 + 24$, i.e., 100.

Here is a step-by-step solution.

---

[22]In what follows we will be referring to isotopes of various atoms. For a discussion on what isotopes are please read Appendix B.

[23]Source: The 1997 report of the IUPAC Subcommittee for Isotopic Abundance Measurements by K.J.R. Rosman, P.D.P. Taylor *Pure Appl. Chem.* **1999**, *71*, 1593-1607. Data have been rounded to the nearest percent.

A Step-by-Step Solution

1. Calculate the mass of the chlorine atoms in a sample of 100 chlorine atoms.

    a. Calculate the mass of ^{35}Cl isotopes in the sample.

    $$35 \times 76 = 2660$$

    The ^{35}Cl isotopes have a mass of 2660 amu.

    b. Calculate the mass of ^{37}Cl isotopes in the sample.

    $$37 \times 24 = 888$$

    The ^{37}Cl isotopes have a mass of 888 amu.

    c. Calculate the mass of the chlorine atoms in the sample.

    $$2660 + 888 = 3548$$

    The chlorine atoms have a mass of 3548 amu.

2. Calculate total number of chlorine atoms in the sample.

    $$76 + 24 = 100$$

    There are 100 chlorine atoms in the sample.

3. Calculate the atomic mass of chlorine.

    $$3548 \div 100 = 35.48$$

    The atomic mass of chlorine is 35.48 amu/atom.

Here is an algebraic solution.[24]

An Algebraic Solution

$$m_a \qquad \text{atomic mass of chlorine (amu/atom)}$$

$$m_a = \frac{35 \times 76 + 37 \times 24}{76 + 24}$$

$$m_a = \frac{2660 + 888}{100}$$

$$m_a = \frac{3548}{100}$$

$$m_a = 35.48$$

The atomic mass of chlorine is 35.48 amu/atom.

---

[24]For an alternative model please see Chapter 7.

The notation m_a is commonly used to represent the value of the quantity *atomic mass*.

Exercise Set 2.3

Solve each of the following problems using the step-by-step and algebraic techniques.

1. Over the past 5 days the temperature was 12°C, 15°C, 18°C, 14°C, and 20°C. What was the average temperature over the 5 days?

2. Your expenses over the past 3 months were as follows: $2100, $1600, and $1950. Based on these, what is your average monthly expense?

3. You ordered 8 Type A computer systems at $1200/system, a Type B computer system for $820, as well as 3 Type C computer systems at $700/system. What is the average cost of a computer system in the order?

4. Last week you worked 20 hours at a job earning $17/hour and 12 hours at another job earning $21/hour. What is your average hourly rate of income?

5. Last week you worked 15 hours at a job earning $14/hour, 10 hours at another job earning $16/hour, and 2 hours at a third job earning $18/hour. What is your average hourly rate of income?

6. There are 120 workers in Section A, 250 workers in Section B, and 320 workers in Section C. Workers in Section A receive 3 hours of training per week. Workers in Section B receive 5 hours of training per week. Workers in Section C receive 8 hours of training per week. What is the average training time per week per worker?

7. At this company, there are 56 workers in Section A, 120 workers in Section B, and 98 workers in Section C. The salaries of workers in Section A, Section B and Section C are $64 000/year, $45 000/year and $52 000/year. What is the average salary of a worker at this company?

8. This term Jorge took three semester-long courses: MATH101, BIOL101 and CHEM101 with credit values of 3, 5 and 4 respectively. He received a B+ in MATH101, corresponding to a grade point value of 3.33, a C− in BIOL101, corresponding to a grade point value of 1.67, and a D in CHEM101, corresponding to a grade point value of 1.00. What is Jorge's term GPA this term?

9. Last term you took PHOT1148, PSY1040, HUM1014 and COMM1017 each of which was one semester long. The contact hours for the four courses were 3 h/week, 2 h/week, 4 h/week and 3 h/week respectively. You received an A in PHOT1148, corresponding to a grade point value of 4.00, a B− in PSY1040, corresponding to a grade point value of 2.67, a C+ in HUM1014, corresponding to a grade point value of 2.33, and an A− in COMM1017, corresponding to a grade point value of 3.67. What was your term GPA last term?

10. A sample of sulfur from nature contains 94.93% ^{32}S, 0.76% ^{33}S, 4.29% ^{34}S, and 0.02% ^{36}S. Calculate the atomic mass of sulfur.[25]

11. A sample of silver from nature contains 51.839% ^{107}Ag and 48.161% ^{109}Ag. Calculate the atomic mass of silver.[26]

12. A sample of oxygen atoms in a sample from the lab contains 64.2% ^{16}O and 35.8% ^{17}O. Calculate the average mass of an atom of oxygen in this sample.

13. A sample of nitrogen atoms in a sample from the lab contains 82% ^{14}N, 13% ^{16}N and 5% ^{17}N. Calculate the average mass of an atom of nitrogen in this sample.

---

[25]Source: The 1997 report of the IUPAC Subcommittee for Isotopic Abundance Measurements by K.J.R. Rosman, P.D.P. Taylor *Pure Appl. Chem.* **1999**, *71*, 1593-1607.

[26]Source: The 1997 report of the IUPAC Subcommittee for Isotopic Abundance Measurements by K.J.R. Rosman, P.D.P. Taylor *Pure Appl. Chem.* **1999**, *71*, 1593-1607.

Part II
Superposition

Superposition refers to the activity of adding and/or subtracting the values of certain quantities to arrive at a net value. In some cases, superposition is the only activity that is needed to solve a problem. In most cases, however, superposition is either absent, is found within the last step in the solution process, or within factors that are enclosed within brackets. In this part of the text we will discuss superposition problems with an emphasis on their structural features and semantic implications.

Chapter 3
General Features of Superposition Problems

In this chapter of the textbook we will define superposition problems and discuss their general features.

3.1 Introduction

Superposition refers to the activity of adding and/or subtracting the values of certain quantities to arrive at a net value. Here is an example of a problem whose solution involves superposition.

Problem

You had 1200 vaccination shots Monday morning. 200 shots were used on Monday. 420 shots were used on Tuesday. On Wednesday you received a delivery of 1500 shots and 720 shots were used. How many shots will you have Thursday morning?

An Algebraic Solution

n number of shots that we will have Thursday morning (1)
$n = 1200 - 200 - 420 + 1500 - 720$

$n = 1360$

We will have 1360 shots Thursday morning.

This is a pure superposition problem as additions and subtractions are the only operations that are present in its model. The expression on the right side of the equation that models the problem above, i.e., $1200 - 200 - 420 +$

1500 − 720, is seen as a superposition of the five terms 1200, 200, 420, 1500 and 720.

Often superposition appears as the last activity in the solution process. Many examples of such problems were given in the previous chapters. Here is another.

Problem

Calculate the mass of 1 mol of C_4H_{10}. The molar masses of C and H are 12.01 g/mol and 1.008 g/mol respectively.

An Algebraic Solution

$$m \quad \text{mass of 1 mol of } C_4H_{10} \text{ (g)}$$
$$m = 12.01 \times 4 + 1.008 \times 10$$

$$m = 48.04 + 10.08$$
$$m = 58.12$$

The mass of 1 mol of C_4H_{10} is 58.12 g.

The expression on the right side of the equation that models the problem above, i.e., $12.01 \times 4 + 1.008 \times 10$, is seen as a superposition of the two terms 12.01×4 and 1.008×10.

In other cases superposition may appear within a factor surrounded by brackets. Here is an example.

Problem

Calculate the mass of 9.6 mol of C_4H_{10}. The molar masses of C and H are 12.01 g/mol and 1.008 g/mol respectively.

An Algebraic Solution

$$m \quad \text{mass of 9.6 mol of } C_4H_{10} \text{ (g)}$$
$$m = 9.6 \, (12.01 \times 4 + 1.008 \times 10)$$

$$m = 9.6 \, (48.04 + 10.08)$$
$$m = 9.6 \times 58.12$$
$$m = 557.952$$

The mass of 9.6 mol of C_4H_{10} is 557.952 g.

The expression on the right side of the equation that models the problem above, i.e., $9.6\,(12.01 \times 4 + 1.008 \times 10)$, analyzes into a single term. This term, i.e., $9.6\,(12.01 \times 4 + 1.008 \times 10)$, analyzes into two factors. These are 9.6 and $(12.01 \times 4 + 1.008 \times 10)$. The second factor is a superposition of the two terms 12.01×4 and 1.008×10.

Superposition may appear both inside a factor *and* across the whole solution as the following example shows.

Problem

Calculate the mass of 1 mol of $Ca\,(OH)_2$. The mass of 1 mol of Ca is 40.08 g. The molar masses of O and H are 16.00 g/mol and 1.008 g/mol respectively.

An Algebraic Solution

$$m \quad \text{mass of 1 mol of } Ca\,(OH)_2 \text{ (g)}$$
$$m = 40.08 + 2\,(16.00 + 1.008)$$

$$m = 48.04 + 2 \times 17.008$$
$$m = 48.04 + 34.016$$
$$m = 82.056$$

The mass of 1 mol of $Ca\,(OH)_2$ is 82.056 g.

The expression on the right side of the equation that models the problem above, i.e., $40.08 + 2\,(16.00 + 1.008)$, is seen as a superposition of the two terms 40.08 and $2\,(16.00 + 1.008)$. The second term breaks into the two factors 2 and $(16.00 + 1.008)$. The second factor is a superposition of the two terms 16.00 and 1.008.

Superposition may involve a single term. Here is an example.

Problem

You bought 3 pens at $2.10/pen. What amount of money did you spend?

An Algebraic Solution

a amount of money I spent (\$)
$a = 2.10 \times 3$

$a = 6.30$

I spent \$6.30.

The expression on the right side of the equation that models the problem above, i.e., 2.10×3, is seen as the superposition of the single term 2.10×3.[1]

No matter where it appears, superposition is easy to spot as it either consists of a single term or has the form of a chain addition/subtraction.

Exercise Set 3.1

1. In each case a word problem and its model are given. Identify the occurrences of superposition in the expression on the right side of the equation that models the problem.

    a. Marie bought 3 pens at \$1.99/pen and a notebook for \$12.80. What amount of money did Marie spend?

    a amount of money Marie spent (\$)
    $a = 1.99 \times 3 + 12.80$

    b. Calculate the mass of a molecule of ozone, O_3. The atomic mass of O is 16.00 amu/atom.

    m mass of a molecule of O_3 (amu)
    $m = 16.00 \times 3$

---

[1]It is quite fair to choose to call such a scenario *a superposition of a single term* as opposed to opting to claim that *such expressions do not represent superposition*. The equation that models the problem above in the body of the text, i.e.,

$a = 2.10 \times 3$

may be written as

$a = 0 + 2.10 \times 3$

Consistent with our earlier view we interpret this as

$$\boxed{a} = \boxed{0} + \boxed{2.10 \times 3}$$

which shows that the term 2.10×3 is being added.

Note that this view sets 0 as the start point of *all* superposition problems, i.e., superposition begins with 0 and then adds and subtracts the values of any quantities that need to be added and subtracted.

c. Calculate the mass of 1 mol of H_2O in g. The molar mass of H is 1.008 g/mol. The mass of 1 mol of O is 16.00 g.

m mass of 1 mol of H_2O (g)
$m = 1.008 \times 2 + 16.00$

d. Stephan had $32.80. He bought 2 pens at $1.30/pen and 4 notebooks at $3.50/notebook. What amount of money does Stephan have left?

a amount of money Stephan has left ($)
$a = 32.80 - (1.30 \times 2 + 3.50 \times 4)$

e. We can extract 4 Cal/g of carbohydrate, 9 Cal/g of fat and 4 Cal/g of protein. A cup of yogurt contains 21.6 g of carbohydrate, 12 g of fat and 10.2 g of protein. What is the energy content of the cup of yogurt?

E energy content of the cup of yogurt (Cal)
$E = 4 \times 21.6 + 9 \times 12 + 4 \times 10.2$

f. You started with 32.4 L of a salt solution. This morning you used the salt solution at a rate of 1.25 L/h for 2.5 h. This afternoon you used the salt solution at a rate of 3.25 L/h for 1.5 h. What volume of the solution do you have left?

V volume of salt solution left (L)
$V = 32.4 - (1.25 \times 2.5 + 3.25 \times 1.5)$

g. Calculate the mass of 4.5 mol of C_2H_4 in g. The molar masses of C and H are 12.01 g/mol and 1.008 g/mol respectively.

m mass of 4.5 mol of C_2H_4 (g)
$m = 4.5\,(12.01 \times 2 + 1.008 \times 4)$

h. This morning you drove at a speed of 120 km/h for 1.5 h. This afternoon you drove at a speed of 90 km/h for 2.5 h. What total distance did you cover today?

d total distance I covered today (km)
$d = 120 \times 1.5 + 90 \times 2.5$

i. Mike had a board with a length of 1.82 m. He cut 3 pieces from the board each of which was 0.35 m long. What is the length of the board that is left after the cuts?

l length of board after the cuts (m)
$l = 1.82 - 0.35 \times 3$

j. There are 2 boxes each of which has a mass of 20.5 g. There are 3 boxes each of which has a mass of 8.5 g. What is the average mass of a box?

$\overline{m}$ average mass of a box (g/box)

$$\overline{m} = \frac{20.5 \times 2 \; + \; 8.5 \times 3}{2 \; + \; 3}$$

k. A sample of O atoms contains 97.3% ^{16}O, 2.6% ^{17}O and 0.1% ^{18}O. Calculate the average mass of an O atom in the sample.

$\overline{m}$ average mass of an O atom in the sample (amu/atom)
$$\overline{m} = 16 \times 0.973 \; + \; 17 \times 0.026 \; + \; 18 \times 0.001$$

l. This morning you drove at a speed of 110 km/h for 1.5 h. This afternoon you drove at a speed of 85 km/h for 3 h. What was your average speed today?

$\overline{v}$ my average speed today (km/h)

$$\overline{v} = \frac{110 \times 1.5 \; + \; 85 \times 3}{1.5 \; + \; 3}$$

3.2   The Place of Superposition in Analysis and Synthesis

Giving priority to additions and subtractions as the first operations along which expressions break down is a choice, not a requirement.[2] The reason we have chosen to analyze expressions along the lines of additions and subtractions first is that superposition problems are the simplest types of problems that we, as humans, solve.

The first question in analyzing expressions, then, is to ask *what entities, i.e., terms, are being added and subtracted in that expression?* We encourage the reader to always look at the terms before analyzing an expression further even in the case when the expression consists of a single term. And, as we will show in the next section, the semantics behind an expression becomes quite clear once we learn how to name the quantities whose values are represented by the terms.

Since additions and subtractions are the first operations that we need to handle during the analysis stage, they rank as last during the synthesis stage which retraces the steps that were taken during the analysis stage. Indeed,

---

[2]It is possible to develop alternate syntaxes that assign priority to other operations. Such alternate syntaxes would require that we decode expressions differently but note that any such alternate syntax would be expected to carry the same semantics.

during synthesis, we first work out the factors, multiply the factors to evaluate the terms, and then add and subtract the terms to arrive at a value for the expression.

Superposition describes the relationship between the subexpressions of a multi-term expression.[3] As an example, with t_1, t_2 and t_3 representing terms, the expression

$$t_1 + t_2 - t_3$$

is superpositional to each of the terms t_1, t_2 and t_3. The relationship may be direct or inverse depending on whether the term is being added or subtracted: The expression $t_1 + t_2 - t_3$ is directly superpositional to each of the terms t_1 and t_2,[4] and inversely superpositional to the term t_3.[5]

Further to the above, we can also say that the expression $t_1 + t_2 - t_3$ is superpositional to any combination of the terms in that expression. As an example, the expression $t_1 + t_2 - t_3$ is superpositional to $t_1 + t_2$.[6] We can show this to be the case by noting that if the sum of t_1 and t_2, i.e., $t_1 + t_2$, increases/decreases by a certain amount, then the value of the expression $t_1 + t_2 - t_3$ will increase/decrease by the same amount.[7] A similar argument shows that the expression $t_1 + t_2 - t_3$ is superpositional to $t_1 - t_3$, $t_2 - t_3$, and $t_1 + t_2 - t_3$, i.e., the expression itself.

---

[3]A **Subexpression** of an expression consists of any combination of the terms within that expression or their paraphrases. As an example, the subexpressions of the expression $t_1 + t_2 - t_3$ are the terms t_1, t_2, t_3, any combination involving two terms such as $t_1 + t_2$, $t_1 - t_3$, and $t_2 - t_3$, and so on up to the combination that involves all the terms, i.e., the expression $t_1 + t_2 - t_3$ itself. Examples of paraphrases of such subexpressions are $t_2 + t_1$ which is a paraphrase of $t_1 + t_2$, $-(t_3 - t_2)$ which is a paraphrase of $t_2 - t_3$, and the like.

[4]This means that an increase in the value of t_1 or t_2 by a certain amount forces an increase in the value of the expression that they are a part of, i.e., $t_1 + t_2 - t_3$, by the same amount and a decrease in the value of t_1 or t_2 by a certain amount forces a decrease in the value of the expression that they are a part of, i.e., $t_1 + t_2 - t_3$, by the same amount.

[5]This means that an increase in the value of t_3 by a certain amount forces a decrease in the value of the expression that it is a part of, i.e., $t_1 + t_2 - t_3$, by the same amount and a decrease in the value of t_3 by a certain amount forces an increase in the value of the expression that it is a part of, i.e., $t_1 + t_2 - t_3$, by the same amount.

[6]Indeed, directly so.

[7]We can also show this to be true if we write the expression $t_1 + t_2 - t_3$ as

$$(t_1 + t_2) - t_3$$

which, in effect, turns the two separate terms, t_1 and t_2, into a single term. The expression $(t_1 + t_2) - t_3$ analyzes into two terms: $(t_1 + t_2)$ and t_3. Since an expression is superpositional to the terms that it analyzes into, the expression $(t_1 + t_2) - t_3$ is superpositional to the term $(t_1 + t_2)$.

Consider now the equation

$$t = t_1 + t_2 - t_3$$

In this equation t is set equal to the expression $t_1+t_2-t_3$. This means that t and $t_1+t_2-t_3$ are one and the same thing.[8] This implies that any conclusions made about the expression $t_1 + t_2 - t_3$ is valid for t as well. As an example, as we have shown, the expression $t_1 + t_2 - t_3$ is directly superpositional to t_1. This immediately implies that t is directly superpositional to t_1 as t is the same as $t_1+t_2-t_3$ and, therefore, any conclusions made about the expression $t_1 + t_2 - t_3$ is valid for t as well.

This observation allows us to relate t to subexpressions of the expression $t_1 + t_2 - t_3$ in the same way that we related $t_1 + t_2 - t_3$ to its subexpressions. In particular, t is directly superpositional to t_1 and t_2 and is inversely superpositional to t_3. In addition, t is directly superpositional to $t_1 + t_2$, $t_1 - t_3$, and the like.

Finally, we note that both direct and inverse superposition are symmetric: If t_2 is directly superpositional to t_1, then t_1 is directly superpositional to t_2,[9] and if t_2 is inversely superpositional to t_1, then t_1 is inversely superpositional to t_2.[10] Justification for these claims will be given in the chapters ahead.

Exercise Set 3.2

1. For each problem in Exercise Set 3.1, Question 1, analyze the expression on the right side of the given model along the lines of superposition.

2. Consider the expression

$$t_1 - t_2 + t_3 - t_4$$

   where t_1, t_2, t_3 and t_4 represent terms.

   a. List the subexpressions of $t_1 - t_2 + t_3 - t_4$ to which $t_1 - t_2 + t_3 - t_4$ is directly superpositional.
   b. List the subexpressions of $t_1 - t_2 + t_3 - t_4$ to which $t_1 - t_2 + t_3 - t_4$ is inversely superpositional.

---

[8] One may think of t as another way to refer to $t_1 + t_2 - t_3$.
[9] This means that, if an increase/decrease in t_1 by a certain amount forces an increase/decrease in t_2 by the same amount, then an increase/decrease in t_2 by a certain amount forces an increase/decrease in t_1 by the same amount.
[10] This means that, if an increase/decrease in t_1 by a certain amount forces a decrease/increase in t_2 by the same amount, then an increase/decrease in t_2 by a certain amount forces a decrease/increase in t_1 by the same amount.

3. Consider the following equation

$$t = -t_1 + t_2 - t_3$$

where t, t_1, t_2 and t_3 represent terms.

   a. What is the relationship between t and t_1?
   b. What is the relationship between t and t_2?
   c. What is the relationship between t and t_3?
   d. What happens to t if t_1 increases by a certain amount?
   e. What happens to t if t_1 decreases by a certain amount?
   f. What happens to t if t_2 increases by a certain amount?
   g. What happens to t if t_2 decreases by a certain amount?
   h. What happens to t if t_3 increases by a certain amount?
   i. What happens to t if t_3 decreases by a certain amount?

4. In what follows q_1 and q_2 represent the values of two quantities.

   a. Suppose that an increase/decrease in q_1 by a certain amount forces an increase/decrease in q_2 by the same amount. What will happen to q_1 if q_2 increases/decreases by a certain amount? Justify your answer.
   b. Suppose that an increase/decrease in q_1 by a certain amount forces a decrease/increase in q_2 by the same amount. What will happen to q_1 if q_2 increases/decreases by a certain amount? Justify your answer.

3.3   The Quantity whose Value is Represented by a Term

In an algebraic equation, every quantity symbol and every numerical value represents the value of a quantity. As an example, consider the following problem.

Problem

You bought 2 pens at \$1.50/pen and 3 notebooks at \$4.99/notebook. What amount of money did you spend in total?

An Algebraic Model

a amount of money I spent (\$)
$a = 1.50 \times 2 + 4.99 \times 3$

This model involves the quantity symbol, a, and the numerical values 1.50, 2, 4.99 and 3. The following relates these entities to the quantities whose values they represent.

a amount of money I spent ($\$$)
1.50   cost of a pen ($\$/1$)
2   number of pens (1)
4.99   cost of a notebook ($\$/1$)
3   number of notebooks (1)

In addition to the quantity symbols and numerical values, every subexpression of the expressions on either side of the equation, and in particular each term within these expressions, represents the value of a certain quantity.

As an example, consider the terms in the equation that models the problem above. On the left side of this equation there is the single term, a, which, as noted above, represents the value of the quantity *amount of money I spent*. On the right side of this equation there are two terms: The first is 1.5×2 which represents the *amount of money spent on the pens*, and the second is 4.99×3 which represents the *amount of money spent on the notebooks*. These terms and the associated quantities whose values they represent are listed below.

a amount of money I spent ($\$$)
1.50×2 amount of money spent on the pens ($\$$)
4.99×3 amount of money spent on the notebooks ($\$$)

The ability to name the quantities whose values are represented by the terms in an equation or an expression helps us make sense of that equation or expression. As an example, with our focus on the terms in the equation that models the problem above, the interpretations that are given for the terms above allow us to read the equation

$$a = 1.50 \times 2 + 4.99 \times 3$$

as

> The *amount of money I spent* is equal to the sum of the *amount of money spent on the pens* and *amount of money spent on the notebooks*.

It should be noted that the quantity that a term represents is different from the quantities that the quantity symbols and numerical values within that term represent. As an example, the first term on the right side of the equation above, i.e., 1.50×2, contains the numerical values 1.50 and 2. However, the quantity whose value is represented by 1.50×2, i.e., the *amount of money spent on the pens*, is different from the quantities whose values are represented by the numerical values 1.50, i.e., the *cost of a pen*, and 2, i.e., the *number of pens*. This is illustrated below.[11,12]

2   number of pens (1)
1.50   cost of a pen ($/1)
1.50×2 amount of money spent on the pens ($)

Problem

For breakfast Roberto had 2 cups of coffee at 145 Cal/cup, a slice of toast with jam containing 180 Cal and 2 eggs at 210 Cal/egg. What is Roberto's net intake of energy from the breakfast?

An Algebraic Model

E Roberto's net intake of energy from breakfast (Cal)
$E = 145 \times 2 + 180 + 210 \times 2$

The equation above involves four terms. These terms and the associated quantities whose values they represent are listed below.

E Roberto's net intake of energy from breakfast (Cal)
145×2 energy intake from the coffee (Cal)
180   energy intake from the toast and jam (Cal)
210×2 energy intake from the eggs (Cal)

---

[11]We have an eye here on the kinds of science formulas that will come our way in the pages ahead. As an example, in the following formula from physics

$$d = d_0 + vt$$

the second term on the right, i.e., vt, represents the distance that an object moving at a speed of v and duration t covers. Note that the quantity whose value is represented by the term vt, i.e., *distance covered*, is very different from the quantities whose values are represented by the quantity symbols v and t, i.e., *speed* and *travel time*.

[12]To figure out the quantity whose value is represented by a term, it may help to evaluate the term and then try to associate a quantity to the value that one arrives at. As an example, to determine the quantity whose value is represented by the term 1.50×2 in the example above, we start by evaluating the term to arrive at 3.00. This represents the amount of money spent on the pens. This is the quantity whose value is represented by the term 1.50×2. Note that *amount of money spent on the pens* is different from *cost of a pen times the number of pens* which describes the manner in which 3.00 is calculated.

With our focus on the semantics behind the terms in the equation that models the problem above, the interpretations given for the terms allow us to read the equation

$$E = 145 \times 2 + 180 + 210 \times 2$$

as

Roberto's net intake of energy from breakfast is equal to the sum of the *energy intake from the coffee, energy intake from the toast and jam* and *energy intake from the eggs.*

Problem

Calculate the mass of 1 mol of $Ca(OH)_2$. The mass of 1 mol of Ca is 40.08 g. The molar masses of O and H are 16.00 g/mol and 1.008 g/mol respectively.

An Algebraic Model

m mass of 1 mol of $Ca(OH)_2$ (g)

$$m = 40.08 + 2(16.00 + 1.008)$$

The equation above involves three terms. On the left side there is the single term m which represents the value of the mass of 1 mol of $Ca(OH)_2$. On the right side there are two terms: 40.08 which represents the value of the mass of 1 mol of Ca, and $2(16.00 + 1.008)$ which represents the value of the mass of 2 mol of OH.[13]

The expression within brackets breaks down into two terms. The first term, i.e., 16.00, represents the molar mass of O. The second term, i.e., 1.008, represents the molar mass of H.

In addition to the numerical values, quantity symbols and terms in an expression, we can associate meaning to subexpressions that consist of a combination of two or more terms. As an example of this, consider the following problem.

---

[13] We let the reader interpret the equation that models the problem above in English.

Problem

This morning Martin had 3.8 L of a solution. He used the solution at a rate of 0.57 L/h for 2.5 h. At noon he received a delivery of 5.5 L of the solution. Later he used 1.8 L of the solution. What volume of the solution does he have left?

An Algebraic Model

$$V \quad \text{volume of solution Martin has left (L)}$$
$$V = 3.8 - 0.57 \times 2.5 + 5.5 - 1.8$$

The quantities whose values are represented by the terms in the expression on the right side of the equation that models the problem are listed below.

3.8	volume of the solution this morning
0.57×2.5	volume of the solution used this morning
5.5	volume of the solution delivered at noon
1.8	volume of the solution used this afternoon

We can combine any number of these terms and speak of the quantity whose value is represented by such a combination. As an example the subexpression made up of the first two terms in the expression $3.8 - 0.57 \times 2.5 + 5.5 - 1.8$, i.e.,

$$3.8 - 0.57 \times 2.5$$

represents the volume of the solution that Martin had after the usage in the morning and before the delivery at noon. Similarly, the subexpression made up of the first three terms in the expression $3.8 - 0.57 \times 2.5 + 5.5 - 1.8$, i.e.,

$$3.8 - 0.57 \times 2.5 + 5.5$$

represents the volume of solution that Martin had after the delivery. Such subexpressions do not need to include the first term and the terms that are combined by them do not need to form a chain. As an example, the subexpression made up of the second and fourth terms of the expression $3.8 - 0.57 \times 2.5 + 5.5 - 1.8$, i.e.,

$$-0.57 \times 2.5 - 1.8$$

represents the total volume of the solution that was used today.

Exercise Set 3.3

1. For each problem in Exercise Set 3.1, Question 1, analyze the given model along the lines of superposition and express the relationship between the terms in the equation that models the problem in English.

2. Consider the following problem.

   > This morning Maria had 2 cups of coffee at 130 Cal/cup and a slice of buttered toast containing 180 Cal. She was inactive for an hour, burning 85 Cal. She then went running and burnt 450 Cal. Following this she had 3 eggs at 210 Cal/egg. What is Maria's net intake of energy so far?

   The following equation models the problem.

   $$E \quad \text{Maria's net intake of energy so far (Cal)}$$
   $$E = 130 \times 2 + 180 - 85 - 450 + 210 \times 3$$

   Name the quantity whose value is represented by each of the following syntactic components of the expression on the right side of the equation that models the problem above.

   a. 130

   b. 2

   c. 180

   d. 85

   e. 450

   f. 210

   g. 3

   h. 130×2

   i. 210×3

   j. $130 \times 2 + 180$

   k. $130 \times 2 + 180 - 85$

   l. $130 \times 2 + 180 - 85 - 450$

   m. $-85 - 450$

   n. $130 \times 2 + 180 + 210 \times 3$

3.4   Sources and Sinks

In an expression that involves the superposition of terms, some terms are added while others are subtracted. Terms that are added increase the value of the expression of which they are a part of and, for this reason, are referred to as **sources** of the quantity whose value is represented by the whole expression. In addition, terms that are subtracted decrease the value of the expression of which they are a part of and, for this reason, are referred to as **sinks** of the quantity whose value is represented by the whole expression. Let us illustrate matters using an example.

Problem

You had 1200 mL of a medication. You used 3 doses at 10.5 mL/dose. You received a shipment of 500 mL of the medication. You used 2 doses at 12.5 mL/dose. What volume of medication do you have left?

An Algebraic Solution

V volume of medication left (mL)
$$V = 1200 - 10.5 \times 3 + 500 - 12.5 \times 2$$

The expression on the right side of the equation that models the problem represents the volume of the medication that is left[14] and is a superposition of four terms. The first term, i.e., 1200, represents the volume of medication at the start and is seen as a source of the volume of medication that is left as it *increases* the volume of medication that is left. This is followed by the term 10.5×3 which represents the volume of medication that was used for the initial set of doses and is seen as a sink of the volume of the medication that is left as it *decreases* the volume of medication that is left. The third term, i.e., 500, represents the volume of medication that was shipped and is seen as a source of the volume of the medication that is left as it *increases* the volume of medication that is left. The last term, i.e., 12.5×2, represents the volume of medication that was used for the second set of doses and is seen as a sink of the volume of the medication that is left as it *decreases* the volume of medication that is left.

To determine whether a term is a source or a sink of the expression of which it is a part of, look at the operation on the left side of that term. If there is a + on the left side of the term, the term is being added and represents a source. If, on the other hand, there is a − on the left side of the term, the term is being subtracted and represents a sink.

---

[14]Note that it is more proper to refer to this as *the value of the volume of the medication that is left* as opposed to *the volume of the medication that is left*. However, in the interest of efficiency, the phrase *value of* is often dropped in such contexts and is used only when one wishes to emphasize the fact that it is not the quantity itself, but its *value* that is under investigation. In fact, if one wanted to be overly proper, one would always use the phrase *value of* in the opening statement of algebraic models as well. As an example, for the problem above, we would write

V the value of the volume of medication left (mL)

as opposed to

V the volume of medication left (mL)

In this textbook we follow the common convention and drop the use of the phrase *value of* unless we wish to emphasize the fact or every now and then remind the reader of the fact.

Problem

Last week Akua worked for 14.5 h at her first job earning \$15.25/h and 8.5 h at her second job earning \$18.10/h. What was Akua's average rate of pay last week?

An Algebraic Model

r Akua's average rate of pay last week (\$/h)

$$r = \frac{15.25 \times 14.5 + 18.10 \times 8.5}{14.5 + 8.5}$$

The left and right side expressions in the equation above involve one term each. The term on the left, i.e., r, represents Akua's average rate of pay last week. The term on the right, i.e., $\frac{15.25 \times 14.5 + 18.10 \times 8.5}{14.5 + 8.5}$, also represents Akua's average rate of pay last week which is, obviously, a source of Akua's average rate of pay last week.

The numerator of the expression on the right side breaks down into two terms. The first term, i.e., 15.25×14.5, represents the amount of money Akua earned last week from her first job and is a source of the expression $15.25 \times 14.5 + 18.10 \times 8.5$ which represents the total amount of money that she earned last week. The second term, i.e., 18.10×8.5, represent the amount of money that she earned from her second job last week and is a source of the expression $15.25 \times 14.5 + 18.10 \times 8.5$ which represents the total amount of money that she made last week.

The denominator of the expression on the right side also breaks into two terms. The first term, i.e., 14.5, represents the duration of work at Akua's first job last week and is a source of the expression $14.5 + 8.5$ which represents the total time Akua worked last week. The second term, i.e., 8.5, represents the duration of work at Akua's second job last week and is also a source of the expression $14.5 + 8.5$ which represents the total time Akua worked last week.

Exercise Set 3.4

1. For each problem in Exercise Set 3.1, Question 1, identify the terms in the expression on the right side of the equation that models the problem and, for each term, identify the quantity whose value is represented by the term, label the term as a source or a sink and name the quantity that it is a source or a sink of.

3.5   Consistency

Terms that have the same units can be added and subtracted but it makes no sense to add or subtract terms that have different units. As an example, it makes sense to add 14.2 kg and 16.7 kg but it does not make sense to add 15.9 kg and 20.5 km.[15]

An equation is **consistent** if all the terms in that equation have the same units.

Problem

Francesca is driving from Toronto to Montreal, a distance of 504 km. This morning she drove at a speed of 120 km/h for 2.5 h. What distance remains to be covered?

The following equation models the problem above.

$$d \quad \text{remaining distance (km)}$$
$$d = 504 - 120 \times 2.5$$

Show that the equation is consistent.

Solution

We list the terms and work out the unit for each term:

Term: d
Unit: km

Term: 504
Unit: km

---

[15]If a mix of different units relating to the same quantity are present, then it is possible to add the terms by converting one of the units to the other. As an example, we can add 4.5 m and 120 cm by converting 4.5 m to 450 cm and then add this value to 120 cm to arrive at a total length of 570 cm. Or we can covert 120 cm to 1.2 m and then proceed to add this to 4.5 m to arrive at a total length of 5.7 m.

By convention, mixed units may not be used in the communication of the values of quantities in the sciences. This is a requirement of the International System of Units, SI, which is the default measurement system used in the sciences. While it may be permissible in other measurement systems to use mixed units (as in quoting one's height as 5 feet 8 inches), in the sciences it is required that all such mixed expressions should be converted to a single value using a single unit through the conversion of the units involved into one common unit followed by the addition and subtraction of the terms involved to generate a value for the quantity that uses a single unit.

Term: 120×2.5
Unit: $\text{km/h} \times \text{h} = \text{km}$

Since all the terms in the equation have the same unit, the equation is consistent.

The unit of the first term listed above, i.e., d, is given by the line

d remaining distance (km)

The unit of the second term listed above, i.e., 504, is given in the body of the word problem as km.

The unit of the third term listed above, i.e., 120×2.5, may be worked out by replacing the numerical values 120 and 2.5 with their respective units km/h and h and then multiplying the units to arrive at km. This is shown below.[16]

$$\text{km/h} \times \text{h} = \frac{\text{km}}{\not{h}} \times \not{h}$$
$$= \text{km}$$

Problem

Calculate the mass of a molecule of C_2H_4. The atomic masses of C and H are 12.01 amu/atom and 1.008 amu/atom respectively.

The following equation models the problem above.

m mass of a molecule of C_2H_4 (amu)
$m = 12.01 \times 2 + 1.008 \times 4$

Show that the equation is consistent.

Solution

We list the terms and work out the unit for each term:

Term: m
Unit: amu

---

[16]With some notable differences, unit symbols can be operated on in much the same way as numerical values as they represent the *sizes* of units.

Term: 12.01×2
Unit:[17] amu$/1 \times 1 =$ amu

Term: 1.008×8
Unit:[18] amu$/1 \times 1 =$ amu

Since all the terms in the equation have the same unit, the equation is consistent.[19]

Exercise Set 3.5

1. For each problem in Exercise Set 3.1, show that the given model is consistent.

3.6   Change in the Value of a Quantity

Temperature at a given location changes as the day progresses. The height of an individual changes throughout their lives. The amount of propane left in the tank changes as we use it to barbecue.

By **change in the value of a quantity** we mean the difference between the final and initial values of that quantity. As an example, if the temperature was 19 °C this morning and it increased to 24 °C at noon, then the change in the value of the temperature is equal to $24 - 19$ or 5 °C. Note that if we started with a temperature of 24 °C and ended with a temperature of 19 °C, then the change in the value of the temperature would have been $19 - 24$ or -5 °C with the negative sign indicating a drop in the value of the temperature.

We use the Greek capital letter delta, Δ, for the phrase *change in* so that Δt is interpreted as *change in the value of temperature* and Δp is interpreted as *change in the value of pressure*.

In general, a positive value of Δq implies a rise in the value of the quantity q. This makes sense as Δq is defined as the difference between the final and initial values of the quantity. If the final value of the quantity is larger than the initial value of the quantity, a scenario that corresponds to a rise in the value of the quantity, then the difference between them will be positive. The following example illustrates the idea.

---

[17]Or, less formally, amu/atom$\times$atoms = amu.

[18]Or, less formally, amu/atom$\times$atoms = amu.

[19]It should be noted that consistency only tells us that, as far as the units of the terms are concerned, the equation makes sense. A consistent equation may still be incorrect on other grounds. As an example, the model $m = 12.01 \times 3 + 1.008 \times 7$ could be consistent but is the wrong model for the mass of a molecule of C_2H_4.

Problem

This morning the temperature was 19 °C. It rose to 24 °C by noon. Calculate the change in the value of the temperature.

An Algebraic Solution

Δt change in the value of temperature (°C)
$\Delta t = 24 - 19$

$\Delta t = 5 °C$

The temperature changed by 5 °C or the temperature rose by 5 °C.

Similarly, a negative value for Δq implies a drop in the value of the quantity q. This, too, makes sense as Δq is defined as the difference between the final and initial values of the quantity. If the final value of the quantity is less than the initial value of the quantity, a scenario that corresponds to a drop in the value of the quantity, then the difference between them will be negative. The following example illustrates the idea.

Problem

This afternoon the temperature was 24 °C. By evening the temperature had dropped to 19 °C. Calculate the change in the value of temperature.

An Algebraic Solution

Δt change in the value of temperature (°C)
$\Delta t = 19 - 24$

$\Delta t = -5 °C$

The temperature changed by -5 °C or the temperature dropped by 5 °C.

Note that the change in the value of a quantity does not tell us what the initial or final values of the quantity were. As an example, a change of 5 °C in the value of temperature could correspond to an initial value of 19 °C and a final value of 24 °C or it could correspond to an initial value of 32 °C and a final value of 37 °C.

Note also that the sign of the change in the value of a quantity is not related to the sign in the initial or final values of that quantity. As an example, a change of 5 °C could take place with initial and final values that are positive (e.g., an initial value of 12 °C and a final value of 17 °C), initial and final values with different signs (e.g., an initial value of −2 °C and a final value of 3 °C) or initial and final values that are negative (e.g., an initial value of −6 °C and a final value of −1 °C).

Exercise Set 3.6

1. The pressure was 320 000 Pa and now measures 323 000 Pa. Calculate the change in the value of pressure.

2. This morning Simon was 120 km from Toronto. This afternoon Simon was 345 km from Toronto. Calculate the change in the value of Simon's distance from Toronto.

3. The temperature was 2.5 °C and dropped to −6.5 °C. Calculate the change in the value of temperature.

4. A balloon with a volume of 8420 cm^3 expands to a volume of 9110 cm^3. What is the change in the value of the volume of the balloon?

5. 4.75 mol of helium is inside a container. The container develops a leak and after some time we find that there is only 3.5 mol of helium left in the container. What is the change in the value of the amount of helium in the container?

6. The temperature was −15.8 °C and rose to −5.9 °C. Calculate the change in the value of temperature.

7. The gravitational potential energy of an object changes from −2400 J to 1350 J. What is the change in the value of the gravitational potential energy of this object?

8. The current through the wire was measured as 0.083 A and now measures 0.055 A. Calculate the change in the value of the current through the wire.

9. a. What does a positive value of Δq imply about the change in q?
   b. What does a negative value of Δq imply about the change in q?

10. If you know Δq, can you work out the initial and final values of q?

11. a. Do the initial and/or final values of q have to be positive if Δq is positive?
    b. Do the initial and/or final values of q have to be negative if Δq is negative?

3.7   Conservation Laws

Conservation laws are statements about the constancy of the result of sums and differences of terms.

Problem

> We can extract 4 Cal/g of carbohydrate, 9 Cal/g of fat and 4 Cal/g of protein. A cup of yogurt contains 4.5 g of carbohydrate, 5.2 g of fat and 8.3 g of protein. Calculate the energy content of the cup of yogurt.

The following equation models the problem.

E energy content of the cup of yogurt (Cal)
$$E = 4 \times 4.5 + 9 \times 5.2 + 4 \times 8.3$$

This model states that the energy content of the cup of yogurt is equal to the sum of energy contributions from the carbohydrate, fat and protein contents of the cup of yogurt.

Note that E is directly superpositional to the terms on the right side of the equation. As an example, an increase of, say, 5 Cal to the first term through the addition of more carbohydrates to the cup of yogurt would result in an increase of 5 Cal in the value of E.

If we were to evaluate the expression on the right side of the equation, we would arrive at a value of 98 Cal as the energy content of the cup of yogurt. Suppose we wish to keep this total the same, or, in other words, to conserve the energy content of the cup of yogurt, but we also wish to increase the value of the first term on the right side of the equation by increasing the amount of carbohydrates in the cup of yogurt. In order to achieve both objectives we can compensate for the additional contribution to the energy content of the cup of yogurt from extra carbohydrates by lowering the contribution to the energy content of the cup of yogurt from other sources. These other sources are represented by the second and the third terms which represent the energy contributions due to the fat and protein contents of the cup of yogurt respectively. Lowering the value of the second term would require reducing the fat content of the cup of yogurt and lowering the value of the third term would require reducing the protein content of the cup of yogurt.

Our next problem is from physics.

In physics, the mechanical energy of an object moving within the gravitational field of another is defined as the sum of its kinetic and potential

energies. The kinetic energy of an object is the energy that an object pos-
sesses due to its state of motion. The potential energy of an object is the
energy that is in some sense stored in that object due to the altitude at which
it is located. It can be shown that the mechanical energy of such an object
is conserved, i.e., its value remains constant as the object moves. With this
in mind, consider the following problem.

Problem

A 5.2 kg projectile is moving at a speed of 0.72 m/s at an altitude
of 12.9 m above the surface of the Earth. Explain the interplay
between the kinetic and potential energies of the object as it falls.
Acceleration due to gravity near the surface of the Earth is approximately 9.8 m/s^2.

The following equation models the problem.

E mechanical energy of the object (J)

$$E = \frac{1}{2} \times 5.2 \times 0.72^2 + 5.2 \times 9.8 \times 12.9$$

The first term in the expression on the right side of the equation above, i.e.,
$\frac{1}{2} \times 5.2 \times 0.72^2$, represents the kinetic energy of the object and the second term
in the expression on the right side of the equation that models the problem
above, i.e., $5.2 \times 9.8 \times 12.9$, represents the potential energy of the object.

As the projectile in the problem above falls, it gains speed. This increases
the value of the first term by a certain amount which increases the value
of E by the same amount as E is directly superpositional to the first term.
However, as stated earlier, according to the laws of physics, the value of the
mechanical energy of an object, i.e., E in the equation above, moving within
the gravitational field of another object is constant, a property that is often
referred to as the Law of Conservation of Mechanical Energy. The only way
for E to remain constant as the value of the first term in the expression on
the right side increases, is for the value of the second term in the expression
on the right side to decrease by an equal amount. This can happen through
a reduction in the altitude at which the object is located which is indeed the
case: As the object falls, it loses altitude.

It is, of course, obvious that as an object falls, it gains speed as its altitude
decreases with the gain in speed raising the value of the first term and the
loss in altitude reducing the value of the second term. What is not so obvious
is that the rise in the value of the first term (i.e., the kinetic energy of the
object) is *equal to* the fall in the value of the second term (i.e., the potential
energy of the object), keeping the sum, i.e., E, constant.

Exercise Set 3.7

1. Consider the following problem.

   > Francesca has a fixed net annual income. She spends $54 000
   > and saves $12 000 of her net annual income per year. What
   > is Francesca's net annual income?

   The following equation models the problem.

   > a Francesca's net annual income ($/year)
   > $a = 54\,000 + 12\,000$

   What happens to Francesca's annual savings if her annual expenses
   increase by $1400/year?

2. Consider the following problem.

   > You are travelling from Toronto to Montreal. This morning
   > you drove at a speed of 120 km/h for 2.5 h. You need to
   > cover another 300 km to reach Montreal. What is the distance
   > between Toronto and Montreal?

   The following equation models the problem.

   > d distance from Toronto to Montreal (km)
   > $d = 120 \times 2.5 + 300$

   How could you have reduced the remaining distance?

3. Consider the following problem.

   > An isolated system consists of two balls moving along a straight
   > line with momenta 5.3 kg·m/s and 4.7 kg·m/s. What is the
   > total momentum of the system?

   The following equation models the problem.

   > p momentum of the system $(kg \cdot m/s)$
   > $p = 5.3 + 4.7$

   The balls collide and, after collision, the momentum of the first ball is
   found to have decreased by 0.12 kg·m/s. What has happened to the
   momentum of the second ball? The Law of Conservation of Momentum
   states that the momentum of an isolated system is conserved.

3.8   Superposition of Percentages

Percentages may be added and/or subtracted if they refer to parts of the same quantity.[20] Consider, as an example, the following problem.

Problem

> 25% of the budget is given to the Research Department and 15% of the budget is given to the Testing Department. What percentage of the budget is given to both the Research and Testing Departments?

In this problem the two percentages 25% and 15% refer to parts of the same quantity, namely, the amount of the budget. As a result, it makes sense to add the percentages to find out the percentage of the budget that is given to both the Research and Testing Departments. This is illustrated in Figure 3.8.1.

The understanding above leads to the following algebraic solution for the problem.

An Algebraic Solution

r percentage of budget given to Research and
Testing Departments (1)

$r = 25\% + 15\%$

$r = 40\%$

40% of the budget is given to Research and Testing Departments.

---

[20]In this section we focus on problems where the relevant sections of the whole are disjoint. As an example of a problem in which the relevant sections of the whole have a nonempty intersection, suppose 20% of the members take tennis lessons, 18% of the members take badminton lessons, and 5% of the members take both tennis lessons and badminton lessons. Under such a scenario, the percentage of the members that take tennis lessons and/or badminton lessons would equal to $20\% + 18\% - 5\%$ or 33% while the percentage that take either tennis or badminton lessons would equal to $20\% + 18\% - 2 \times 5\%$ or 28%. While we do not discuss such scenarios in this textbook, we do wish to point out that, even under such scenarios, the main operations in the relevant model are those that relate to superposition, i.e., additions and subtractions.

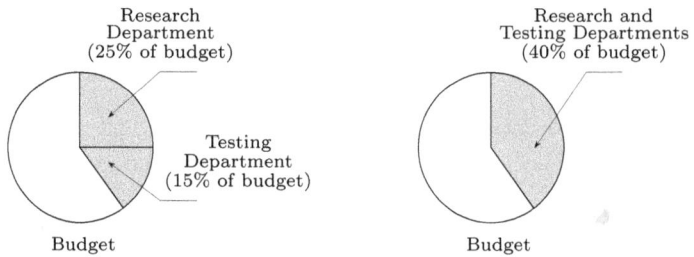

(a) Percentage of the budget given to each of the Research and Testing Departments

(b) Percentage of the budget given to both the Research and Testing Departments

Figure 3.8.1: Percentage of the budget given to both the Research and Testing departments as the sum of the percentage of the budget given to the Research Department and the percentage of the budget given to the Testing Department

Note once again that the reason we can add the percentages in the problem above is that they refer to parts of the same quantity, i.e., the amount of the budget. This is illustrated below.

25% **of the budget** is given to the Research Department.

15% **of the budget** is given to the Testing Department.

40% **of the budget** is given to both Departments.

Problem

A book is on sale at 10% off. Tax is added at 13% of the sale price. Eric argues that the net rate of tax is 13% − 10% or 3%. Explain Eric's error.

Solution

Percentages that refer to parts of different quantities may not be added or subtracted as such additions and subtractions are meaningless. This explains why one cannot subtract the rate of discount from the rate of tax as the rate of discount refers to a percentage of the *original price* whereas the rate of tax refers to a percentage of the *sale price*.

For the problem above, we have the following illustration.

10% **of the original price** is the amount of discount.

13% **of the sale price** is the amount of tax.

3% **of ????** ...

As you can see, it makes no sense to subtract the percentages as shown. Subtraction of 10% from 13% yields 3%, but 3% of what?

Exercise Set 3.8

1. Which of the following percentages can be added and subtracted? Justify your response.

   a. 33.3% of the school budget and 45% of the school budget

   b. 30% of daycare centres in Toronto and 14% of daycare centres in Toronto

   c. 18% of the tomatoes and 14% of the bananas

   d. 12.8% of the workforce and 53.2% of the workforce

   e. 20% of the price and 13% of the sale price

   f. 82% of the volume of the acid solution and 8% of the volume of the acid solution

   g. 17% of the mass of the salt and 3% of the mass of the sugar

   h. 19% of the students in Section A and 22% of the students in Section B

   i. 42% of my income and 36% of my income

   j. 73.5% of the eligible voters and 17% of those who voted

   k. 78% of the turtles in the sample and 14% of the turtles in the sample

2. Solve each of the following problems using the algebraic technique.

   a. 6% of the students got a grade in the A range. 17% of the students got a grade in the B range. What percentage of the students got at least a B−?

   b. 38% of the students got at least a B−. 12% of the students got at least an A−. What percentage of the students got a grade in the B range?

    c. You spend 40% of your income on rent, 25% of your income on groceries and 15% of your income on entertainment. What percentage of your income is left after these expenses?

    d. 73% of the employees working at the ABC plant are full-time. What percentage of the employees at the ABC plant are non-full-time?

    e. Steve completed 40% of the project two weeks ago, 25% of the project last week, and 20% of the project this week. What percentage of the project has been completed?

    f. Maria completed 20% of the project last week, and 10% of the project this week. What percentage of the project has not been completed?

3. In each case determine whether the given argument is valid. Justify your answer.

    a. 3.5% of the employees in Section A and 2.8% of the employees in Section B are away on sick-day leave. Antoine argues that this implies that 3.5% + 2.8% or 6.3% of the total number of employees in Sections A and B are away on sick-day leave.

    b. The balloon developed two leaks. 8.5% of the helium in the balloon was lost through the first leak and 3.7% of the helium in the balloon was lost through the second leak. We can, therefore, conclude that, in total, 8.5% + 3.7% or 12.2% of the helium in the balloon has been lost.

    c. 78% of the crops in the first field and 82% of the crops in the second field have been affected by adverse weather conditions. The report concludes that 82% + 78% or 160% of the crops in both fields have been affected by adverse weather conditions.

    d. 22% of the frogs in the sample have green skin. Of these, 12.8% have red eyes. Therefore, 22% + 12.8% or 34.8% of the frogs in the sample have green skin and red eyes.

    e. 65.8% of the customers are in their 20s and 21.8% of the customers are in their 30s. Maggie says that this implies that 65.8% + 21.8% or 87.6% of the customers are either in their 20s or in their 30s.

    f. 59% of the eligible voters voted. Of those who voted, 15% opposed the proposed changes. Of those who opposed the proposed changes, 3.5% said that they will support a strike if the changes are implemented. Bob says that this means that 59% + 15% + 3.5% or 77.5% of eligible voters who voted will support a strike if the changes are implemented.

Chapter 4
Direct Superposition

If q_1 and q_2 represent the values of two quantities, then we say that q_2 is directly superpositional to q_1 if an increase/decrease in q_1 by a certain amount forces an increase/decrease in q_2 by the same amount.[1]

As an example of relationships of type direct superposition, consider the following problem.

Problem

This morning Steve drove at a speed of 90 km/h for 3.5 h. This afternoon Steve drove at a speed of 120 km/h for 1.5 h. What total distance did Steve cover today?

An Algebraic Solution

d distance Steve covered today (km)
$d = 90 \times 3.5 + 120 \times 1.5$

$d = 315 + 180$
$d = 495$

Steve covered a distance of 495 km today.

---

[1]In the designation *direct superposition* the adjective *direct* is used to imply that the change in q_2 is in the same direction as the change in q_1, i.e., they both increase or they both decrease in value. Such relationships are called **direct relationships**. The word *superposition* implies that the changes take place through additions and subtractions and that the size of the change in q_2 is equal to the size of the change in q_1. Note that q_2 may be directly related to q_1 (in which case they both increase or they both decrease in value) but not be directly superpositional to q_1 (the size of the increase or decrease in one is not equal to the size of the increase or decrease in the other). In a direct superposition relationship, not only do q_1 and q_2 both increase or both decrease in value, but they do so by the same amounts.

Analysis of the equation that models the problem above along the lines of superposition yields three terms. These terms along with the quantities whose values they represent are listed below.

d distance Steve covered today (km)
90×3.5 distance Steve covered this morning (km)
120×1.5 distance Steve covered this afternoon (km)

The equation that models the problem states that the distance that Steve covered today is equal to the sum of the distance that Steve covered this morning and the distance that Steve covered this afternoon.

In this problem the quantity *distance Steve covered today* is directly superpositional to the quantity *distance Steve covered this morning*. Note that any increase/decrease in the value of the distance that Steve covered this morning by a certain amount forces an increase/decrease in the value of the quantity *distance Steve covered today* by the same amount. As an example, in the problem above Steve covered a distance of 90×3.5 or 315 km this morning. This results in a total value of 495 km for the distance Steve covered today. Had Steve driven at, say, 80 km/h this morning, the distance that he would have covered would have been 280 km which is 35 km less than the original. In response to this change, the value of the quantity *distance Steve covered today* goes down by the same amount, i.e., 35 km, to 460 km.

In a given expression, the value of the quantity represented by the expression is directly superpositional to the value of the quantity represented by any of the terms in that expression that are being added, i.e., any terms with a $+$ on their left side. As an example, the expression

$$q_1 + q_2 - q_3 + q_4$$

is directly superpositional to q_1, q_2 and q_4. This makes sense: If we increase the value of q_1, q_2 or q_4 by a certain amount, say 5 units, then we will be adding 5 more and this would result in an increase in the value of the expression $q_1 + q_2 - q_3 + q_4$ by 5 units.

Since the terms that are being added in an expression are sources of the quantity that is represented by the expression, the terms to which the expression is directly superpositional are sources of the quantity that is represented by the expression.

As noted above, the expression

$$q_1 + q_2 - q_3 + q_4$$

is directly superpositional to each of q_1, q_2 and q_4. However, in addition to the above, we can also say that the expression $q_1 + q_2 - q_3 + q_4$ is directly superpositional to any of its subexpressions that includes a subset of the terms q_1, q_2, q_3 and q_4. As an example, the expression $q_1 + q_2 - q_3 + q_4$ is directly superpositional to $q_1 + q_2$. To see how, we write the expression $q_1 + q_2 - q_3 + q_4$ as

$$(q_1 + q_2) - q_3 + q_4$$

which, in effect, turns the two separate terms, q_1 and q_2, into the single term $(q_1 + q_2)$. We now see that an increase/decrease in $(q_1 + q_2)$ by a certain amount forces an increase/decrease in the expression $(q_1 + q_2) - q_3 + q_4$ by the same amount.[2] Similarly, the expression $q_1 + q_2 - q_3 + q_4$ is directly superpositional to the subexpressions $q_1 - q_3$, $q_1 + q_4$, $q_2 - q_3$, $q_2 + q_4$, $-q_3 + q_4$, $q_1 + q_2 - q_3$, $q_1 + q_2 + q_4$, $q_1 - q_3 + q_4$, $q_2 - q_3 + q_4$ and $q_1 + q_2 - q_3 + q_4$.

As discussed earlier, direct superposition is symmetric. If the value of one quantity is directly superpositional to the value of another quantity, then the value of the second quantity is directly superpositional to the value of the first quantity. This means that, if an increase/decrease in the value of one quantity by a certain amount forces an increase/decrease in the value of another quantity by the same amount, then an increase/decrease in the value of the second quantity by a certain amount forces an increase/decrease in the value of the first quantity by the same amount. Justification for this claim will be given in the chapters ahead.

Exercise Set 4

1. Suppose q_1 and q_2 represent the values of two quantities. What does it mean when we say that q_2 *is directly related to* q_1?

2. Suppose q_1 and q_2 represent the values of two quantities. What does it mean when we say that q_2 *is directly superpositional to* q_1?

3. Suppose q_1 and q_2 represent the values of two quantities. Is it possible for q_2 to be directly related to q_1 but not be directly superpositional to q_1?

4. In each case a problem and its model are given. Identify the terms in the expression on the right side of the model to which the value of the expression, i.e., the quantity symbol on the left side, is directly superpositional.

    a. Mina had 32.7 L of a solution and received 3 shipments of the solution at 12.5 L/shipment. What volume of the solution does Mina have

---

[2] Alternatively, we can argue that, since an expression is directly superpositional to the terms in that expression that are added, then the expression $(q_1 + q_2) - q_3 + q_4$ is directly superpositional to the term $(q_1 + q_2)$.

now?

$$V \quad \text{volume of the solution that Mina has now (L)}$$
$$V = 32.7 + 12.5 \times 3$$

b. This morning Marco had 2 slices of toast at 88 Cal/slice and a cup of coffee containing 138 Cal. He then exercised for 1.5 h, burning, on average, 187 Cal/h. Following this, he had 2 eggs at 120 Cal/egg. What was Marco's net intake of energy this morning?

$$E \quad \text{Marco's net intake of energy this morning (Cal)}$$
$$E = 88 \times 2 + 138 - 187 \times 1.5 + 120 \times 2$$

c. This morning the temperature was 15.7 °C. This afternoon the temperature was 18.9 °C. What is the change in the value of the temperature?

$$\Delta t \quad \text{change in the value of temperature (}^\circ\text{C)}$$
$$\Delta t = 18.9 - 15.7$$

d. Enrique has a net annual income of $45 000 and spends $38 500 of this money and saves the rest. How much money does Enrique save in a year?

$$a \quad \text{amount of money Enrique saves every year (\$)}$$
$$a = 45\,000 - 38\,500$$

e. Phillip had $102.56. He bought 2 pens at $2.30/pen and 3 notebooks at $5.99/notebook. What amount of money does Phillip have left?

$$a \quad \text{amount of money Phillip has left (\$)}$$
$$a = 102.56 - (2.30 \times 2 + 5.99 \times 3)$$

f. The plant received 135 deliveries of recycling material at 2.8 t/delivery and has processed 280 t of recycling material so far. What mass of recycling material remains to be processed?

$$m \quad \text{mass of recycling material waiting to be processed (t)}$$
$$m = 2.8 \times 135 - 280$$

g. A swimming pool has a volume of 175 m³ and is being emptied by two pumps. The first removes water at a rate of 35 m³/h and has been working for 0.5 h. The second removes water at a rate of 40 m³/h and has been working for 0.25 h. What volume of water remains to be removed?

$$V \quad \text{volume of water remaining (m}^3\text{)}$$
$$V = 175 - (35 \times 0.5 + 40 \times 0.25)$$

5. For each problem in Question 4 above, identify the terms in the expression on the right side of the equation that are sources of the term on the left side of the equation and describe the relationship between the quantities that are represented by these terms.

6. List the subexpressions of the expression

$$-q_1 + q_2 - q_3 + q_4$$

to which it is directly superpositional.

Chapter 5
Inverse Superposition

If q_1 and q_2 represent the values of two quantities, then we say that q_2 is inversely superpositional to q_1 if an increase/decrease in q_1 by a certain amount forces a decrease/increase in q_2 by the same amount.[1]

As an example of relationships of type inverse superposition, consider the following problem.

Problem

Miranda makes \$65 000 per year, spends 52 000 every year and saves the rest. What amount of money does Miranda save every year?

An Algebraic Solution

a amount of money Miranda saves every year (\$)
$a = 65\,000 - 52\,000$

$a = 13\,000$

Miranda saves \$13 000 every year.

---

[1] In the designation *inverse superposition* the adjective *inverse* is used to imply that the change in q_2 is in a direction opposite that of change in q_1, i.e., one increases in value while the other decreases in value. Such relationships are called **inverse relationships**. The word *superposition* implies that the changes take place through additions and subtractions and that the size of the change in q_2 is equal to the size of the change in q_1. Note that q_2 may be inversely related to q_1 (in which case one increases in value while the other decreases in value) but not be inversely superpositional to q_1 (the size of the increase or decrease in one is not equal to the size of the decrease or increase in the other). In an inverse superposition relationship, not only does q_1 increase in value and q_2 decrease in value (or vice versa), but the size of the increase in one is equal to the size of the decrease in the other.

Analysis of the equation that models the problem above along the lines of superposition yields three terms. These terms along with the quantities whose values they represent are listed below.

a amount of money Miranda saves every year ($)
65 000   amount of money Miranda makes every year ($)
52 000   amount of money Miranda spends every year ($)

The equation that models the problem states that the amount of money Miranda saves every year is equal to the difference between the amount of money that she makes every year and the amount of money that she spends every year.

In this problem the quantity *amount of money Miranda saves every year* is inversely superpositional to the quantity *amount of money Miranda spends every year*. Note that any increase/decrease in the value of the quantity *amount of money Miranda spends every year* by a certain amount forces a decrease/increase in the value of the quantity *amount of money Miranda saves every year* by the same amount. As an example, in the problem above Miranda spends $52 000 every year. This results in savings of $13 000 every year. If Miranda were to spend an additional $1000 every year, then the amount of money that she saves every year decreases by the same amount, i.e., $1000.

In a given expression, the value of the quantity represented by the expression is inversely superpositional to the value of the quantity represented by any of the terms in that expression that are being subtracted, i.e., any terms with a − on their left side. As an example, the expression

$$-q_1 + q_2 - q_3 - q_4$$

is inversely superpositional to q_1, q_3 and q_4. This makes sense: If we increase the value of q_1, q_3 or q_4 by a certain amount, say 5 units, then we will be subtracting 5 more and this would result in a decrease in the value of the expression $-q_1 + q_2 - q_3 - q_4$ by 5 units.

Since the terms that are being subtracted in an expression are sinks of the quantity that is represented by the expression, the terms to which the expression is inversely superpositional are sinks of the quantity that is represented by the expression.

As noted above, the expression

$$-q_1 + q_2 - q_3 - q_4$$

is inversely superpositional to each of q_1, q_3 and q_4. However, in addition to the above, we can also say that the expression $-q_1 + q_2 - q_3 - q_4$ is inversely superpositional to any of its subexpressions that includes a subset of the terms q_1, q_2, q_3 and q_4 so long as a factor of -1 is taken out of the subexpression. As an example, the expression $-q_1 + q_2 - q_3 - q_4$ is inversely superpositional to $(q_1 - q_2)$. To see how, we write the expression $-q_1 + q_2 - q_3 - q_4$ as

$$- (q_1 - q_2) - q_3 - q_4$$

which, in effect, turns the two separate terms, q_1 and q_2, into a single term with a factor of -1 pulled out of the subexpression. We now see that an increase/decrease in $(q_1 - q_2)$ by a certain amount forces a decrease/increase in the expression $- (q_1 - q_2) - q_3 - q_4$ by the same amount.[2] Similarly, the expression $-q_1 + q_2 - q_3 - q_4$ is inversely superpositional to the subexpressions $q_1 + q_3$, $q_1 + q_4$, $-q_2 + q_3$, $-q_2 + q_4$, $q_3 + q_4$, $q_1 - q_2 + q_3$, $q_1 - q_2 + q_4$, $q_1 + q_3 + q_4$, $-q_2 + q_3 + q_4$ and $q_1 - q_2 + q_3 + q_4$.

As discussed earlier, inverse superposition is symmetric. If the value of one quantity is inversely superpositional to the value of another quantity, then the value of the second quantity is inversely superpositional to the value of the first quantity. This means that, if an increase/decrease in the value of one quantity by a certain amount forces a decrease/increase in the value of another quantity by the same amount, then an increase/decrease in the value of the second quantity by a certain amount forces a decrease/increase in the value of the first quantity by the same amount. Justification for this claim will be given in the chapters ahead.

Exercise Set 5

1. Suppose q_1 and q_2 represent the values of two quantities. What does it mean when we say that q_2 *is inversely related to* q_1?

2. Suppose q_1 and q_2 represent the values of two quantities. What does it mean when we say that q_2 *is inversely superpositional to* q_1?

3. Suppose q_1 and q_2 represent the values of two quantities. Is it possible for q_2 to be inversely related to q_1 but not be inversely superpositional to q_1?

4. For each problem in Question 4 in Exercise Set 4, identify the terms in the expression on the right side of the model to which the value of the expression, i.e., the quantity symbol on the left side, is inversely superpositional.

---

[2] Alternatively, we can argue that, since an expression is inversely superpositional to the terms in that expression that are subtracted, then the expression $- (q_1 - q_2) - q_3 - q_4$ is inversely superpositional to the term $(q_1 - q_2)$.

5. For each problem in Question 4 in Exercise Set 4, identify the terms in the expression on the right side of the equation that are sinks of the term on the left side of the equation and describe the relationship between the quantities that are represented by the terms.

6. List the subexpressions of the expression

$$-q_1 + q_2 - q_3 + q_4$$

to which it is inversely superpositional.

Part III
Proportion

Proportion refers to the activity of multiplying and/or dividing the values of certain quantities to arrive at a net value. In some cases, proportion is the only activity that is needed to solve a problem. In most cases, however, proportion is either absent or is found within the terms or arguments of other operations. In this part of the text we will discuss proportion problems with an emphasis on their structural features and semantic implications.

Chapter 6
General Features of Proportion Problems

In this chapter of the textbook we will define proportion problems and discuss their general features.

6.1 Introduction

Proportion relates to the activity of multiplying and/or dividing the values of certain quantities to arrive at a value. In this section we will study the general features of proportion problems starting with single-term expressions that involve multiplication, followed by single-term expressions that involve division, followed by single-term expressions that involve both multiplication and division. The section ends with a discussion on proportion in the context of multi-term expressions.

Single-Term Expressions Involving Multiplication

Consider the following problem.

Problem

You bought 7 pens at $1.20/pen. What amount of money did you spend in total?

Here is an algebraic solution.

An Algebraic Solution

a amount of money I spent ($)
$a = 1.20 \times 7$

$a = 8.40$

In total I spent $8.40.

The expression on the right side of the equation that models the problem, i.e., 1.20×7, analyzes into the single term 1.20×7 which further analyzes into the two factor 1.20 and 7. The term 1.20×7 is seen as a proportion problem involving the two factors 1.20 and 7.

Problem

Jamie bought 3.5 L of a solution at $12.25/L. Later he bought 4.25 L of the same solution at the same price. What amount of money did Jamie spend in total?

An Algebraic Solution

a amount of money Jamie spent ($)
$a = 12.25\,(3.5 + 4.25)$

$a = 12.25 \times 7.75$

$a = 94.94$

In total Jamie spent $94.94.

The expression on the right side of the equation that models the problem, i.e., $12.25\,(3.5 + 4.25)$, analyzes into the single term $12.25\,(3.5 + 4.25)$ which further analyzes into the two factors 12.25 and $(3.5 + 4.25)$. As such, we see the term as a proportion involving the two factors 12.25 and $(3.5 + 4.25)$.

The following problem is an example of a proportion problem involving three factors.

Problem

Meena ordered 75 boxes of printer paper. Each box contains 8 packages of printer paper. Each package contains 120 sheets of printer paper. How many sheets of printer paper did Meena order?

An Algebraic Solution

n number of printer papers Meena ordered (1)
$n = 120 \times 8 \times 75$

$n = 72\,000$

Meena ordered 72 000 sheets of printer paper.

The expression on the right side of the equation that models the problem, i.e., $120 \times 8 \times 75$, analyzes into the single term $120 \times 8 \times 75$ which analyzes further into the three factors 120, 8 and 75. As such, we see the term as a proportion involving the three factors 120, 8 and 75.

Single-Term Expressions Involving Division

Consider the following problem.

Problem

Maria had 140 g of a substance and divided this into 4 equal parts. What is the mass of each part?

An Algebraic Solution

m mass of each part (g)
$m = \dfrac{140}{4}$

$m = 35$

Each part has a mass of 35 g.

The expression on the right side of the equation that models the problem, i.e., $\frac{140}{4}$, analyzes into the single term $\frac{140}{4}$ which analyzes further into the single factor $\frac{140}{4}$. When a factor involves division, we may choose to further analyze the division into factors in its dividend and factors in its divisor. This allows us to view the expression as a proportion involving the factors in the dividend and the divisor of the division. For the expression $\frac{140}{4}$, the dividend analyzes into the single term 140 which analyzes into the single factor 140 and the divisor analyzes into the single term 4 which analyzes into the single factor 4. This allows us to view the term $\frac{140}{4}$ as a proportion involving the factors 140 and 4.[1]

An alternative model for the problem above would argue that the mass of each part is $\frac{1}{4}$ of the total mass.[2] This leads to the following alternative algebraic solution.

An Alternative Algebraic Solution

m mass of each part (g)

$$m = \frac{1}{4} \times 140$$

$$m = 35$$

Each part has a mass of 35 g.

The expression on the right side of the equation that models the problem, i.e., $\frac{1}{4} \times 140$, in the alternative formulation above analyzes into the single term $\frac{1}{4} \times 140$ which consists of the two factors $\frac{1}{4}$ and 140. As such, we may see this as a proportion problem involving the two factors $\frac{1}{4}$ and 140. However, as discussed earlier, we may also view this as a proportion involving the three factors 1, 4, and 140.

Problem

4.5 mol of CO_2 has a mass of 198.045 g. What is the molar mass of CO_2?

---

[1] Note that neither 140 nor 4 is a factor of $\frac{140}{4}$.

[2] Both lines of reasoning are commonly used and the reader should be aware of their equivalence. As an example, $\frac{18}{3}$ (18 divided into 3 parts) is the same as $\frac{1}{3} \times 18$ ($\frac{1}{3}$ of 18), $\frac{15}{4}$ (15 divided by 4) is the same as $\frac{1}{4} \times 15$ ($\frac{1}{4}$ of 15).

The reading of multiplication as *of* in such contexts mimics the use of the word in dealing with whole numbers. Note that by 2 *of something* we mean 2× the thing and by 3 *of something* we mean 3× the thing. Similarly, *half of something* maps onto $\frac{1}{2} \times$ the thing and $\frac{1}{3}$ *of something* maps onto $\frac{1}{3} \times$ the thing. The multiplication algorithm captures such semantics. As an example, $\frac{1}{3}$ of 6 is 2 as is $\frac{1}{3} \times 6$.

An Algebraic Solution

M molar mass of CO_2 (g/mol)

$$M = \frac{198.045}{4.5}$$

$$M = 44.01$$

The molar mass of CO_2 is 44.01 g/mol.

The expression on the right side of the equation that models the problem above, i.e., $\frac{198.045}{4.5}$, analyzes into the single term $\frac{198.045}{4.5}$ which analyzes into the single factor $\frac{198.045}{4.5}$. This factor can be analyzed further into the dividend 198.045, which consists of a single term containing the single factor 198.045, and the divisor 4.5, which consists of a single term containing the single factor 4.5. The expression is seen as a proportion involving the two factors 198.045 and 4.5.

Single-Term Expressions Involving Multiplication and Division

Consider the following problem.

Problem

5 pens cost $6.00. At this rate, what is the cost of 7 pens?

An Algebraic Solution

c cost of the pens ($)

$$c = \frac{6.00}{5} \times 7$$

$$c = 8.40$$

The pens cost $8.40.

The expression on the right side of the equation that models the problem above,[3] i.e., $\frac{6.00}{5} \times 7$, analyzes into a single term, i.e., $\frac{6.00}{5} \times 7$. The single

---

[3]The reader may be familiar with solution techniques that use the following setup.

$$\frac{6.00}{5} = \frac{c}{7}$$

term analyzes into the two factors $\frac{6.00}{5}$ and 7. We may, therefore, see this expression as a proportion involving the two factors $\frac{6.00}{5}$ and 7. As before, we may choose to analyze the first factor further into the dividend, 6.00, which analyzes into a single term containing the single factor 6.00, and the divisor, 5, which analyzes into a single term containing the single factor 5. This view allows us to see the expression as a proportion involving the three factors 6.00, 5 and 7.

Problem

Jane had 14.7 L of a salt solution. She used $\frac{2}{3}$ of this volume. What volume of the salt solution did Jane use?

An Algebraic Solution

V volume of salt solution used (L)

$$V = \frac{2}{3} \times 14.7$$

$$V = 9.8$$

Jane used 9.8 L of the salt solution.

We let the reader analyze the expression on the right side of the model above along the lines of proportion.

Multi-Term Expressions

Often proportion appears as the last activity in the solution process. This is the case in single-term expressions as illustrated by the examples given above. In other cases proportion may appear within terms in multi-term expressions. Here is an example.

Such setups, along with others, are discussed in detail in Appendices D and E where we present the logic behind such alternative setups and compare them to the setups that are presented in the body of the text. Here we simply note that such setups, while capable of generating algebraic models for single proportion problems, are unable to do the same for problems whose solutions involve chain proportion without resorting to the use of systems. For this reason, such alternatives are not covered in the body of the text.

Problem

Calculate the mass of 1 mol of C_4H_{10}. The molar masses of C and H are 12.01 g/mol and 1.008 g/mol respectively.

An Algebraic Solution

m mass of 1 mol of C_4H_{10} (g)
$m = 12.01 \times 4 + 1.008 \times 10$

$m = 48.04 + 10.08$
$m = 58.12$

The mass of 1 mol of C_4H_{10} is 58.12 g.

The expression on the right side of the equation that models the problem above, i.e., $12.01 \times 4 + 1.008 \times 10$, is a superposition of the two terms 12.01×4 and 1.008×10. The first term is a proportion involving the two factors 12.01 and 4. The second term is a proportion involving the two factors 1.008 and 10.

Proportion may appear both inside a term *and* across the whole solution as the following example shows.

Problem

Calculate the mass of 9.6 mol of C_4H_{10}. The molar masses of C and H are 12.01 g/mol and 1.008 g/mol respectively.

An Algebraic Solution

m mass of 9.6 mol of C_4H_{10} (g)
$m = 9.6\,(12.01 \times 4 + 1.008 \times 10)$

$m = 9.6\,(48.04 + 10.08)$
$m = 9.6 \times 58.12$
$m = 557.952$

The mass of 9.6 mol of C_4H_{10} is 557.952 g.

The expression on the right side of the equation that models the problem analyzes into the single term $9.6\,(12.01 \times 4 + 1.008 \times 10)$ which is a proportion problem involving the factors 9.6 and $(12.01 \times 4 + 1.008 \times 10)$. The second factor is a superposition of the two terms 12.01×4, which is a proportion involving the two factors 12.01 and 4, and 1.008×10, which is a proportion involving the two factors 1.008 and 10.

In some cases proportion may involve a single factor. Here is an example.

Problem

Calculate the mass of a molecule of H_2O. The molar mass of H is 1.008 amu/atom. The mass of an atom of O is 16.00 amu.

An Algebraic Solution

m mass of a molecule of H_2O (amu)
$m = 1.008 \times 2 + 16.00$

$m = 2.016 + 16$
$m = 18.016$

The mass of a molecule of H_2O is 18.016 amu.

The expression on the right side of the equation that models the problem above, i.e., $1.008 \times 2 + 16.00$, is seen as a superposition of the two terms 1.008×2, which is a proportion involving the two factors 1.008 and 2, and 16.00, which is a proportion involving the single factor 16.00.[4]

When a factor in an expression involves division, one may further analyze the division into factors in its dividend and factors in its divisor. We presented an example of this kind at the beginning of this section. Here is another.

Problem

A right circular prism has a base area of 4.2 cm^2 and a volume of 9.3 cm^3. Calculate the prism's height.

---

[4]It is quite fair to choose to call such a scenario a *proportion involving a single factor* as opposed to opting to claim that *such expressions do not represent proportion*. The term involving the single factor, i.e.,

16.00

may be written as

1×16.00

which shows that the factor 16.00 is being multiplied by.

An Algebraic Solution

h the prism's height (cm)

$$h = \frac{9.3}{4.2}$$

$$h = 2.21$$

The prism has a height of 2.21 cm.

The expression on the right side of the equation that models the problem, i.e., $\frac{9.3}{4.2}$, analyzes into the single term, $\frac{9.3}{4.2}$, which further analyzes into the single factor $\frac{9.3}{4.2}$. This allows us to conclude that the expression is a proportion involving the single factor $\frac{9.3}{4.2}$.

If we choose to do so, we can analyze this factor further. This factor involves a division with the single term 9.3 as dividend which analyzes into the single factor 9.3, and the single term 4.2 as divisor which further analyzes into the single factor 4.2. If we choose to do so, we may say that the expression is a proportion involving the factors in its dividend, i.e., 9.3, and divisor, i.e., 4.2.

No matter where it appears, proportion is easy to spot as it either consists of a single factor or has the form of a chain multiplication/division appearing within terms.

Exercise Set 6.1

1. In each case a word problem and its model are given. Identify the occurrences of proportion in the expression on the right side of the equation that models the problem.

   a. Sheeva bought 2 pens at \$0.87/pen and 3 notebook at \$7.20/notebook. What amount of money did Sheeva spend?

   a amount of money Sheeva spent (\$)
   $a = 0.87 \times 2 + 7.20 \times 3$

   b. Calculate the mass of 1 mol of hydrogen, H_2. The molar mass of H is 1.008 g/mol.

   m mass of 1 mol of H_2 (g)
   $m = 1.008 \times 2$

   c. Calculate the mass of a molecule of CO_2 in amu. The mass of an atom of C is 12.01 amu. The atomic mass of O is 16.00 amu/atom.

   m mass of a molecule of CO_2 (amu)
   $m = 12.01 + 16.00 \times 2$

d. You had \$41.83. He bought 3 pens at \$1.10/pen and 4 notebooks at \$6.25/notebook. What amount of money do you have left?

a amount of money you have left (\$)
$a = 41.83 - (1.10 \times 3 + 6.25 \times 4)$

e. We can extract 4 Cal/g of carbohydrate, 9 Cal/g of fat and 4 Cal/g of protein. A cup of yogurt contains 11.4 g of carbohydrate, 4.5 g of fat and 7.8 g of protein. What is its energy content of the cup of yogurt?

E energy content of the cup of yogurt (Cal)
$E = 4 \times 11.4 + 9 \times 4.5 + 4 \times 7.8$

f. You started with 41.8 L of a salt solution. This morning you used the salt solution at a rate of 2.1 L/h for 1.5 h. This afternoon you used the salt solution at a rate of 4.7 L/h for 2 h. What volume of the salt solution do you have left?

V volume of salt solution left (L)
$V = 41.8 - (2.1 \times 1.5 + 4.7 \times 2)$

g. Calculate the mass of 2.3 mol of C_3H_8 in g. The molar masses of C and H are 12.01 g/mol and 1.008 g/mol respectively.

m mass of the C_3H_8 (g)
$m = 2.3 (12.01 \times 3 + 1.008 \times 8)$

h. This morning you drove at a speed of 110 km/h for 1.5 h. This afternoon you drove at a speed of 80 km/h for 2.5 h. What total distance did you cover today?

d total distance I covered today (km)
$d = 110 \times 1.5 + 80 \times 2.5$

i. There are 3 boxes each of which has a mass of 11.9 g. There are 2 boxes each of which has a mass of 7.1 g. What is the average mass of a box?

$\overline{m}$ average mass of a box (g/box)
$$\overline{m} = \frac{11.9 \times 3 + 7.1 \times 2}{3 + 2}$$

j. A sample of O atoms contains 97.3% ^{16}O, 2.6% ^{17}O and 0.1% ^{18}O. Calculate the average mass of an O atom in the sample.

$\overline{m}$ average mass of an O atom in the sample (amu/atom)
$\overline{m} = 16 \times 0.973 + 17 \times 0.026 + 18 \times 0.001$

k. This morning you drove at a speed of 110 km/h for 1.5 h. This afternoon you drove at a speed of 85 km/h for 3 h. What was your average speed today?

$\overline{v}$ my average speed today (km/h)

$$\overline{v} = \frac{110 \times 1.5 \ + \ 85 \times 3}{1.5 \ + \ 3}$$

6.2   The Place of Proportion in Analysis and Synthesis

Analysis of an algebraic expression begins with the splitting of the expression into terms which are entities within an expression that are added and subtracted. This establishes a relationship between the expression and the terms that it analyzes into: An expression is superpositional to each term that it analyzes into.

Each term, in turn, analyzes into factors which are entities within terms that are multiplied and divided.[5] This establishes a relationship between a term and the factors that it analyzes into: A term is proportional to the factors that it analyzes into.

Giving priority to multiplications and divisions as operations along which terms break down is a choice, not a requirement.[6] The reason we have chosen to analyze terms along the lines of multiplications and divisions first is that proportion problems are the next simplest types of problems beyond superposition that we, as humans, solve.

The first question in analyzing terms, then, is to ask *what entities, i.e., factors, are being multiplied and divided in that term?* We encourage the reader to always look at the factors before analyzing a term further even in the case when the term consists of a single factor. And, as we will show in the next section, the semantics behind a term becomes quite clear once we learn how to name the quantities whose values are represented by the factors within the term.

Since multiplications and divisions are the next operations[7] that we need to handle during the analysis stage, they rank as the next-to-last operations that we work out during the synthesis stage which retraces the steps that were taken during the analysis stage. Indeed, during synthesis, we first work out the factors, multiply and divide the factors to evaluate the terms, and then add and subtract the terms to arrive at a value for the expression.

---

[5] As explained earlier, factors that involve division may be further analyzed into factors in the dividend and the divisor in that division.

[6] It is possible to develop alternate syntaxes that assign priority to other operations. Such alternate syntaxes would require that we decode terms differently but note that any such alternate syntax would be expected to carry the same semantics.

[7] Next to additions and subtractions.

Proportion describes the relationship between the subexpressions of a term.[8] As an example, with f_1, f_2 and f_3 representing factors, the expression

$$\frac{f_1 f_2}{f_3}$$

is proportional to each of the factors f_1, f_2 and f_3. The relationship may be direct or inverse depending on whether the factors are being multiplied or divided by: The expression $\frac{f_1 f_2}{f_3}$ is directly proportional to the factors f_1 and f_2,[9] and inversely proportional to the factor f_3.[10]

Further to the above, we can also say that the expression $\frac{f_1 f_2}{f_3}$ is proportional to any combination of the factors in that term. As an example, $\frac{f_1 f_2}{f_3}$ is proportional to $f_1 f_2$.[11]. We can show this to be the case by noting that scaling the product of f_1 and f_2, i.e., $f_1 f_2$, by a certain factor forces the scaling of the expression $\frac{f_1 f_2}{f_3}$ by the same factor.[12] A similar argument shows that the expression $\frac{f_1 f_2}{f_3}$ is proportional to $\frac{f_1}{f_3}$, $\frac{f_2}{f_3}$, and $\frac{f_1 f_2}{f_3}$, i.e., the term itself.

Consider now the equation

$$f = \frac{f_1 f_2}{f_3}$$

In this equation f is set equal to the expression $\frac{f_1 f_2}{f_3}$. This means that f and $\frac{f_1 f_2}{f_3}$ are one and the same thing.[13] This implies that any conclusions made about the term $\frac{f_1 f_2}{f_3}$ is valid for f as well. As an example, as shown above, the term $\frac{f_1 f_2}{f_3}$ is directly proportional to f_1. This immediately implies that f is directly proportional to f_1 as f is the same as $\frac{f_1 f_2}{f_3}$ and, therefore, any conclusions made about the term $\frac{f_1 f_2}{f_3}$ is valid for f as well.

---

[8]A **subexpression** of a term consists of any combination of factors within that term. If a factor involves division, then the subexpression may involve any combination of the factors in its dividend and divisor.

[9]This means that scaling f_1 or f_2 by a certain factor forces the scaling of the term that they are a part of, i.e., $\frac{f_1 f_2}{f_3}$, by the same factor.

[10]This means that scaling f_3 by a certain factor, forces the scaling of the term that it is a part of, i.e., $\frac{f_1 f_2}{f_3}$, by the inverse of the same factor.

[11]indeed, directly so

[12]We can also show this to be true if we write the expression $\frac{f_1 f_2}{f_3}$ as

$$\frac{(f_1 f_2)}{f_3}$$

which, in effect, turns the separate factors, f_1 and f_2, into a single factor $(f_1 f_2)$. Since a term is proportional to the factors that it analyzes into, the term $\frac{(f_1 f_2)}{f_3}$ is proportional to the factor $(f_1 f_2)$.

[13]One may think of f as another way to refer to $\frac{f_1 f_2}{f_3}$.

This observation allows us to relate f to subexpressions of the term $\frac{f_1 f_2}{f_3}$ in the same way that we related $\frac{f_1 f_2}{f_3}$ to its subexpressions. In particular, f is directly proportional to f_1 and f_2 and inversely proportional to f_3. In addition, f is directly proportional to $f_1 f_2$, $\frac{f_1}{f_3}$, and the like.

Multi-term expressions may be analyzed similarly. As an example, consider the multi-term expression

$$f_1 f_2 + \frac{f_3}{f_4}$$

Analysis of this expression yields two terms: $f_1 f_2$ and $\frac{f_3}{f_4}$. This is shown below:

$$\overset{t_1}{\boxed{f_1 f_2}} + \overset{t_2}{\boxed{\dfrac{f_3}{f_4}}}$$

Since each term is proportional to the factors within it, We can make the following assertions:[14] t_1 is directly proportional to f_1 and f_2, and t_2 is directly proportional to f_3 and inversely proportional to f_4.

Finally, we note that both direct and inverse proportion are symmetric: If f_2 is directly proportional to f_1, then f_1 is directly proportional to f_2,[15] and if f_2 is inversely proportional to f_1, then f_1 is inversely proportional to f_2.[16] Justification for these claims will be given in the chapters ahead.

Exercise Set 6.2

1. For each problem in Exercise Set 6.1, analyze the expression on the right side of the given model along the lines of superposition followed by proportion.

2. Consider the term

$$\frac{f_1 f_2}{f_3 f_4}$$

where f_1, f_2, f_3 and f_4 represent factors.

---

[14] We can, of course, make other assertions as well such as the assertion that t_1 is directly proportional to $f_1 f_2$ or the assertion that t_2 is inversely proportional to $\frac{1}{f_3}$ and the like.

[15] This means that, if it is the case that scaling f_1 by a certain factor forces the scaling of f_2 by the same factor, then scaling f_2 by a certain factor forces the scaling of f_1 by the same factor.

[16] This means that, if it is the case that scaling f_1 by a certain factor forces the scaling of f_2 by the inverse of the same factor, then scaling f_2 by a certain factor forces the scaling of f_1 by the inverse of the same factor.

a. List the subexpressions of the term $\frac{f_1 f_2}{f_3 f_4}$ to which $\frac{f_1 f_2}{f_3 f_4}$ is directly proportional.

b. List the subexpressions of the term $\frac{f_1 f_2}{f_3 f_4}$ to which $\frac{f_1 f_2}{f_3 f_4}$ is inversely proportional.

3. Consider the following equation

$$f = \frac{f_1}{f_2 f_3}$$

where f, f_1, f_2 and f_3 represent factors.

a. What is the relationship between f and f_1?

b. What is the relationship between f and f_2?

c. What is the relationship between f and f_3?

d. What happens to f if f_1 is scaled by a certain factor?

e. What happens to f if f_2 is scaled by a certain factor?

f. What happens to f if f_3 is scaled by a certain factor?

g. What happens to f if f_1 is multiplied by a certain value?

h. What happens to f if f_1 is divided by a certain value?

i. What happens to f if f_2 is multiplied by a certain value?

j. What happens to f if f_2 is divided by a certain value?

k. What happens to f if f_3 is multiplied by a certain value?

l. What happens to f if f_3 is divided by a certain value?

4. Consider the following expression

$$\frac{f_1}{f_2 f_3} + \frac{1}{f_4}$$

where f_1, f_2, f_3 and f_4 represent factors.

Analyze the expression into terms and state and interpret the relationship between each term and the factors within that term.

5. In what follows q_1 and q_2 represent the values of two quantities.

a. Suppose that scaling q_1 by a certain factor forces the scaling of q_2 by the same factor. What will happen to q_1 if q_2 is scaled by a certain factor? Justify your answer.

b. Suppose that scaling q_1 by a certain factor forces the scaling of q_2 by the inverse of the same factor. What will happen to q_1 if q_2 is scaled by a certain factor? Justify your answer.

6.3   The Quantity whose Value is Represented by a Factor

As we stated earlier in the text, every term in an expression presents the value of a quantity. We will now show that every factor within a term also represents the value of a quantity. As an example, consider the following problem.

Problem

You bought 2 pens at \$1.50/pen and 3 notebooks at \$4.99/notebook. What amount of money did you spend in total?

An Algebraic Model

a amount of money I spent (\$)
$a = 1.50 \times 2 + 4.99 \times 3$

The expression on the right side of the equation that models the problem, i.e., $1.50 \times 2 + 4.99 \times 3$ analyzes into the two terms 1.50×2 and 4.99×3.

The first term, i.e., 1.50×2, represents the value of the quantity *amount of money spent on the pens*. The term analyzes into the two factors 1.50 which represents the value of *the cost of a pen* and 2 which represents the value of *the number of pens*.

The second term, i.e., 4.99×3, represents the value of the quantity *amount of money spent on the notebooks*. The term analyzes into the two factors 4.99 which represents the value of the quantity *the cost of a notebook* and 3 which represents the value of the quantity *the number of notebooks*.

The ability to name the quantities whose values are represented by the terms and the factors within those terms help us make sense of those terms. As an example, the first term in the expression on the right side of the equation that models the problem above, i.e.,

1.50×2

represents the *amount of money spent on the pens*. This term analyzes into the two factors 1.50, which represents the cost of a pen, and 2, which represents the number of pens purchased. These interpretations allow us to make sense of the term by relating the quantity whose value is represented by the term to the quantities whose values are represented by the factors within that term: The *amount of money spent on the pens* is equal to the product of the *cost of a pen* and *the number of pens purchased*.

Problem

Calculate the mass of 1 mol of $Ca(OH)_2$. The mass of 1 mol of Ca is 40.08 g. The molar masses of O and H are 16.00 g/mol and 1.008 g/mol respectively.

An Algebraic Model

$$m \quad \text{mass of 1 mol of } Ca(OH)_2 \text{ (g)}$$
$$m = 40.08 + 2(16.00 + 1.008)$$

The expression on the right side of the equation that models the problem, i.e., $40.08 + 2(16.00 + 1.008)$, analyzes into two terms.

The first term, i.e., 40.08, consists of the single factor 40.08. This represents the mass of 1 mol of Ca in 1 mol of $Ca(OH)_2$. While it is possible to relate the term 40.08 to the factor 40.08, the relationship is trivial: The mass of 1 mol of Ca (i.e., the term 40.08) is directly proportional to the mass of 1 mol of Ca (i.e., the factor 40.08). This is of course obvious: If we were to scale the factor 40.08 by a factor of 2 as an example (i.e., doubling the factor, 40.08), then the term 40.08 would also scale by a factor of 2 (i.e., the term, 40.08, would double).

The second term, i.e., $2(16.00 + 1.008)$, analyzes into two factors. The first factor, i.e., 2, represents the amount of OH in 1 mol of $Ca(OH)_2$. The second factor, i.e., $(16.00 + 1.008)$, represents the molar mass of OH. These interpretations allow us to make sense of the term by relating the quantity whose value is represented by the term to the quantities whose values are represented by the factors within that term: The mass of the groups of OH in 1 mol of $Ca(OH)_2$ (i.e., the term $2(16.00 + 1.008)$) is equal to twice the molar mass of OH (i.e., the product of the factors 2 which is the amount of OH in 1 mol of $Ca(OH)_2$ and the molar mass of OH).

Factors that involve division may be viewed as a quotient, a dividend and a divisor.

Problem

2 pizzas are to be divided equally among 3 people. How much pizza does each person get?

An Algebraic Model

a amount of pizza each person gets (1)

$$a = \frac{2}{3}$$

The expression on the right side of the model above consists of the single term, $\frac{2}{3}$, that contains the single factor, $\frac{2}{3}$. The factor $\frac{2}{3}$ can be analyzed further into the dividend, 2, which represents the number of pizzas, and the divisor, 3, which represents the number of people.

Exercise Set 6.3

1. For each problem in Exercise Set 6.1, analyze the expression on the right side of the given model along the lines of superposition and proportion and express the relationship between each expression and its terms, as well as each term and its factors in English.

Chapter 7
Direct Proportion

If q_1 and q_2 represent the values of two quantities, then we say that q_2 is directly proportional to q_1 if scaling q_1 by a certain factor[1] forces the scaling of q_2 by the same factor.[2]

As noted in the chapter on direct superposition, in the designation *direct proportion* the adjective *direct* is used to imply that the change in the size of q_2 is in the same direction as the change in the size of q_1, i.e., they both increase or both decrease in size. The word *proportion* implies that the change takes place through multiplications and divisions. Note that a relationship may be direct (in which case the size of the values of the two quantities both increase or both decrease) but not direct proportion (the sizes of the values of the two quantities do not increase or decrease by the same factor). In a relationship of type *direct proportion*, not only do the sizes of q_1 and q_2 both increase or both decrease, but they do so by the same factor.

As an example of a relationship of type direct proportion, consider the following everyday life problem.

Problem

Elizabeth bought 4 pens at \$0.69/pen. What amount of money did Elizabeth spend in total?

---

[1] Scaling a value by a factor means multiplying the value by that factor. As an example, scaling the value of mass by a factor of 2 means doubling the value of the mass and scaling the value of speed by a factor of $\frac{1}{2}$ means halving the value of the speed and so on.

[2] This implies that multiplication/division of the value of one of the quantities by a number forces the multiplication/division of the value of the other quantity by the same number. As an example, multiplication of the value of one of the quantities by 2 (i.e., doubling its value), forces the multiplication of the value of the other quantity by 2 (i.e., doubles the value of the other quantity) and multiplying the value of one of the quantities by $\frac{1}{2}$ (i.e., halving its value or dividing its value by 2) forces the multiplication of the value of the other quantity by $\frac{1}{2}$ (i.e., halves its value or divides its value by 2).

An Algebraic Solution

> a amount of money Elizabeth spent ($)
> $a = 0.69 \times 4$
>
> $a = 2.76$
>
> In total, Elizabeth spent $2.76.

Analysis of the expression on the right side of the equation that models the problem yields the single term 0.69×4, which breaks into the two factors 0.69 and 4. The term, the factors within the term along with the quantities whose values they represent are listed below.

> 0.69×4 amount of money Elizabeth spent ($)
> 0.69 the price of a pen ($/1)
> 4 number of pens purchased (1)

This is illustrated below:

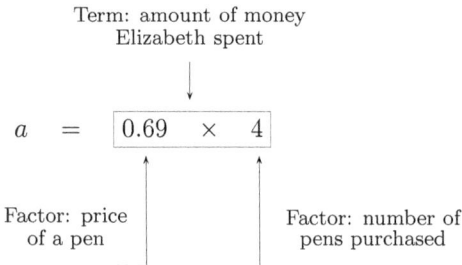

Term: amount of money
Elizabeth spent

$$a \quad = \quad \boxed{0.69 \quad \times \quad 4}$$

Factor: price
of a pen

Factor: number of
pens purchased

The equation that models the problem states that the *amount of money Elizabeth spent* is equal to the product of the *price of a pen* and the *number of pens purchased*.

In this problem the *amount of money Elizabeth spent* is directly proportional to the *price of a pen*. Note that scaling the price of a pen by a certain factor forces the scaling of the amount of money Elizabeth spent by the same factor. As an example, in the problem above Elizabeth spent a total of 0.69×4 or $2.76 on the pens. Had the price of a pen been, say, $1.38, which is twice $0.69, the amount of money Elizabeth spent on the pens would have doubled to a total of 1.38×4 or $5.52.

Similarly, the *amount of money Elizabeth spent* is directly proportional to the *number of pens purchased*. As an example, in the problem above Elizabeth spent a total of 0.69×4 or $2.76 on the pens. Had Elizabeth purchased, say,

8 pens, which is twice 4, then the amount of money Elizabeth spent on the pens would have doubled to a total of 0.69×8 or $5.52.

The quantity represented by a term is always proportional to the quantities that are represented by the factors in it. If the expression is made up of a single term, then the quantity represented by the whole expression is the same as the quantity represented by the single term that it is made of and we can say that the quantity represented by the expression as a whole is proportional to the quantities represented by the factors within its single term. This was the case in the problem above. When the expression is made up of multiple terms, the quantity represented by each term is proportional to the quantities represented by the factors within it but the quantity represented by the whole expression is *not* proportional to any of the factors within any of its terms. The following problem illustrates the point.

Problem

This morning you drove at a speed of 120 km/h for 1.5 h. This afternoon you drove at a speed of 90 km/h for 2.5 h. What total distance did you cover today?

An Algebraic Model

d distance I covered today (km)
$d = 120 \times 1.5 + 90 \times 2.5$

The equation above states that *distance I covered today* (represented by the term d) is equal to the sum of *distance I covered this morning* (represented by the term 120×1.5) and *distance I covered this afternoon* (represented by the term 90×2.5).

Analysis of the expression on the right side of the equation that models the problem into terms and factors yields the following.[3]

---

[3]We invite the reader to name the quantities that are represented by the second term and the factors within the second term.

Term: distance covered
this morning

$$\downarrow$$

$$a \;\; = \;\; \boxed{120 \;\; \times \;\; 1.5} \;\; + \;\; \boxed{90 \;\; \times \;\; 2.5}$$

Factor: speed   Factor: travel time
this morning   this morning

In this problem the *distance I covered this morning* is directly proportional to *my speed this morning*. Note that scaling my speed by a factor of, say, $\frac{1}{2}$ (i.e., driving at half the speed this morning), would result in the scaling of the distance I covered this morning by a factor of $\frac{1}{2}$ (i.e., halves the distance that I covered this morning). Similarly, the *distance I covered this morning* is directly proportional to *travel time this morning*. Note that scaling my travel time this morning by a factor of 2 (i.e., driving twice longer this morning) would result in the scaling of the distance I covered this morning by a factor of 2 (i.e., doubles the distance I covered this morning).

As we just noted, the quantity represented by the first term in the expression on the right side of the equation above, i.e., *distance covered this morning*, is directly proportional to the quantities that the factors within it represent, i.e., *driving speed this morning* and *travel time this morning*. However, the quantity represented by the whole expression, i.e., *distance covered today*, is not directly proportional to the quantities that the factors within the first term represent, i.e., *driving speed this morning* and *travel time this morning*. As an example, while doubling travel time this morning (doubling the factor 1.5) doubles distance covered this morning (doubles the term 120×1.5), it does not double distance covered today (does *not* double the expression $120 \times 1.5 + 90 \times 2.5$).

Let us now look at some of the applications of direct proportion.[4]

Everyday Life Problems

Consider the following problem.

Problem

5 pens cost \$6.00. At this rate, what is the cost of 2 pens?

---

[4]In what follows we will use rates to setup our models. For an alternative approach using ratios please see Appendix D.

We will present an algebraic solution for this problem. Following this, we will explain how the model for this problem is put together and the reason this setup works.

First we note that this is an example of a direct proportion problem since the cost of the pens is directly proportional to the number of pens purchased. As an example, the purchase of twice as many pens would cost twice as much and the purchase of half as many pens would cost half as much. Since the problem is of type *direct proportion*, we can follow the guidelines below to model the problem.

To setup the expression on the right side of the equation that models the problem, it may help to make a note of the correspondence between the two values 5 pens and $6.00 as shown below.

5 pens $6.00

This correspondence is given by the sentence 5 *pens cost* $6.00 in the body of the word problem.

We now begin with the value that is left, i.e., 2 pens, and we write

5 pens $6.00

2 pens

Since there is one relationship listed (i.e., the relationship between 5 pens and $6.00), our expression will involve one division as shown below.

5 pens $6.00

$$\frac{}{} \times 2 \text{ pens}$$

The related values 5 pens and $6.00 will act as the dividend and divisor of this division with the one that has the same unit as the unit of 2 (i.e., 1, or, informally, *pens*) acting as the divisor, i.e.,

5 pens $6.00   ✓

$$\frac{\$6.00}{5 \text{ pens}} \times 2 \text{ pens}$$

Note that we have checked off the relationship between 5 pens and $6.00.[5] We now drop the units to get

$$\frac{6.00}{5} \times 2$$

The full solution without the intervening explanations is given below.

---

[5]This checking off of the used relationship is optional here but becomes necessary when there are multiple relationships listed as the checkmark reminds us that the values in the relationship should no longer be used. See the example that follows.

An Algebraic Solution

c cost of the pens (\$)

$$c = \frac{6.00}{5} \times 2$$

$$c = \$2.40$$

The pens cost \$2.40.

The factor $\frac{6.00}{5}$ is an example of a *rate*. A **rate** relates the values of two quantities with different units through division. In the example above the rate $\frac{6.00}{5}$ tells us that 5 pens cost \$6.00.

Dividing the dividend by the divisor in a rate generates a unit rate. A **unit rate** relates the value of one quantity to a *unit* of some other quantity through division. In the example above, division of 6.00 by 5, yields the unit rate $\frac{1.20}{1}$. This represents the cost of 1 pen and it is for this reason that we refer to it as a *unit* rate.

Unit rates are often communicated using the keyword *per* with the forward slash, i.e., /, as the corresponding mathematical symbol . As an example, the unit rate $\frac{\$1.20}{1 \text{ pen}}$ in the problem above is often written as \$1.20/pen and read as \$1.20 *per* pen. In this sense, *per* means *for one*.

To see how the expression on the right side of the equation that models the problem above solves the problem, we first convert the rate $\frac{6.00}{5}$ to a unit rate through the division of the dividend, 6.00, by the divisor, 5, as discussed above. As illustrated below, this yields \$1.20/pen as the unit rate representing the cost of a pen.

$$\frac{\$6.00}{5 \text{ pens}} \times 2 \text{ pens}$$

$$\downarrow$$

$$\$1.20/\text{pen} \times 2 \text{ pens}$$

We can now multiply the cost of a pen, i.e., \$1.20/pen, by the number of pens, i.e., 2 pens, to get \$2.40 as the cost of the pens. Note that the units *pens* cancel out leaving \$ as the unit of the result.

The line of reasoning in solving the problem above (from the rate $\frac{\$6.00}{5 \text{ pens}}$ to the unit rate \$1.20/pen to the cost of the pens \$1.20/pen×2 pens) may be used in the solution of any direct proportion problem and is the only logic that is needed.

Problem

> 2.5 L of a solution has a mass of 423.8 g. 20 g of the solution costs
> $3.80. At this rate, how much does 13.75 L of this solution cost?

First we note that this is an example of a direct proportion problem, i.e.,
a problem where the sizes of the quantities involved (volume of solution, mass
of solution, and cost of solution) all increase or decrease by the same factor.
As an example, four times as much volume implies four times the mass and
four times the cost. Since this is a direct proportion problem, we can use the
line of reasoning given in the previous problem to solve this problem.

To begin, we note the relationships that are given:

2.5 L   423.8 g
20 g $3.80

We now begin with the value 13.75.[6] We write

2.5 L   423.8 g
20 g $3.80
13.75 L

Since there are two relationships in the list above (one relating 2.5 L to 423.8 g
and the other relating 20 g to $3.80), our expression will involve two divisions
as shown below.

2.5 L   423.8 g
20 g $3.80
——— × ——— × 13.75 L

The pair of numbers in each relationship will appear as the dividend and
divisor of each of the divisions in the model above. To figure out where the
values go, we first note the unit of 13.75, i.e., L. This requires that we place
a value with the unit L as the divisor of the first division on the right. The
only value in the listed relationships that has the unit L, is 2.5. We write

2.5 L   423.8 g ✓
20 g $3.80

——— × $\dfrac{423.8 \text{ g}}{2.5 \text{ L}}$ × 13.75 L

Note that we have checked off the relationship whose values were just used
in populating the first division on the right. This reminds us that the values
in this relationship should no longer be used.

---

[6]It is easy to find this value: It is the value that does not appear in the relationships
that are listed.

The values in the last relationship appear as the dividend and divisor of the remaining division with the value whose unit is the same as the unit of the dividend of the previous division appearing as the divisor of the new division. This gives us

$$\frac{2.5\ \text{L}}{20\ \text{g}} \qquad \frac{423.8\ \text{g}}{\$3.80} \quad \checkmark$$
$$\frac{\$3.80}{20\ \text{g}} \times \frac{423.8\ \text{g}}{2.5\ \text{L}} \times 13.75\ \text{L}$$

To see how the expression above models the problem, we begin by converting the rates into unit rates. This yields the following.

$$\$0.19/\text{g} \times 169.52\ \text{g/L} \times 13.75\ \text{L}$$

Performing the first multiplication on the right yields the following:[7]

$$\$0.19/\text{g} \times 2330.9\ \text{g}$$

Performing the last multiplication yields the following:[8]

$$\$442.87$$

The full solution without the intervening comments and explanations is given below:

An Algebraic Solution

c cost of the solution ($)

$$c = \frac{3.80}{20} \times \frac{423.8}{2.5} \times 13.75$$
$$c = \$442.87$$

The solution costs $442.87.

---

[7] The logic behind this move is identical to the logic behind the move from $2.50/pen × 2 pens to $5: To find the cost of the pens we multiply the cost of a pen by the number of pens. The logic behind the move from 169.52 g/L × 13.75 L to 2330.9 g is similar: To find the mass of the solution, we multiply the mass of one litre of the solution by the volume of the solution in litres.

[8] The logic behind this move is identical to the logic behind the move from $2.50/pen × 2 pens to $5: To find the cost of the pens we multiply the cost of a pen by the number of pens. The logic behind the move from $0.19/g × 2330.9 g to $442.87 is similar: To find the cost of the solution we multiply the cost of one gram of the solution by the mass of the solution in grams.

Conversion of Units

Most units of measure used in measuring the value of a given quantity are directly proportional to each other.[9] As an example, length in metres is directly proportional to length in centimetres. To convert units that are directly proportional to each other we can use the setup given earlier in the solution of everyday life problems.

Problem

A beam is 32.4 in long. Convert this to ft given that

1 ft = 12 in

We will present an algebraic solution for this problem. Following this, we will explain how the model for this problem is put together and the reason this setup works.

To setup the expression on the right side of the equation that models the problem, it may help to make a note of the correspondence between the two values 1 ft and 12 in as shown below.

1 ft   12 in

This correspondence is given in the body of the problem or is otherwise known.

We now begin with the value whose unit we want to convert, i.e., 32.4 in, and we write

1 ft   12 in
32.4 in

---

[9] As this statement implies, there are units of measure that are not directly proportional to each other. Among the more important such units are the units of temperature, degree Celsius, degree Fahrenheit and kelvin. The setups given in the body of the text work only if the units involved are directly proportional to each other and, therefore, should not be used to convert units that do not exhibit such relationship. To check to see whether two units are directly proportional to each other check to see whether (a) the value of 0 for one unit corresponds to the value of 0 for the other unit and (b) if there is a fixed relationship between the sizes of the two units. If either one of these conditions fails, then the relationship between the two units is not direct proportion. As an example, if the two units involved are metres and centimetres, then the relationship between them is direct proportion since 0 m corresponds to 0 cm and there is fixed relationship between metres and centimetres: 1 m = 100 cm. If the two units involved are degree Celsius and degree Fahrenheit, then the relationship between them is not direct proportion as 0 °C corresponds to 32 °F, not 0 °F.

Since there is one relationship listed (i.e., the relationship between 1 ft and 12 in), our expression will involve one division as shown below.

$$\frac{1 \text{ ft} \qquad 12 \text{ in}}{} \times 32.4 \text{ in}$$

The related values 1 ft and 12 in will act as the dividend and divisor of this division with the one that has the same unit as the unit of 32.4 (i.e., in) acting as the divisor, i.e.,

$$1 \text{ ft} \qquad 12 \text{ in} \quad \checkmark \;\cdot$$
$$\frac{1 \text{ ft}}{12 \text{ in}} \times 32.4 \text{ in}$$

Note that we have checked off the relationship between 1 ft and 12 in.[10] We now drop the units to get

$$\frac{1}{12} \times 32.4$$

The full solution without the intervening explanations is given below.

An Algebraic Solution

L length of the beam (ft)

$$L = \frac{1}{12} \times 32.4$$
$$L = 2.7 \text{ ft}$$

The beam has a length of 2.7 ft.

To see how the expression on the right side of the equation that models the problem above solves the problem, we first convert the rate $\frac{1}{12}$ to a unit rate through the division of the dividend, 1, by the divisor, 12. This yields $0.08333\ldots$ ft/in and tells us what fraction of a foot fits into an inch.

$$\frac{1 \text{ ft}}{12 \text{ in}} \times 32.4 \text{ in}$$

$$0.08333\ldots \text{ ft/in} \times 32.4 \text{ in}$$

Now that we know how many feet fit in an inch, we multiply by the length in inches to calculate the length in feet.

---

[10]This checking off of the used relationship is optional here but becomes necessary when there are multiple relationships listed as the checkmark reminds us that the values in the relationship should no longer be used. See the example that follows.

Problem

A beam has a length of 125.4 cm. Convert this to ft given that

$$1 \text{ in} = 2.54 \text{ cm}$$
$$1 \text{ ft} = 12 \text{ in}$$

First we note that this is an example of a direct proportion problem, i.e., a problem where the sizes of the quantities involved (length in centimetres, length in inches and length in feet) all increase or decrease by the same factor. As an example, four times the length in centimetres implies four times the length in inches and four times the length in feet. Since this is a direct proportion problem, we can extend the line of reasoning given in the previous problem to solve this problem.

To begin, we note the relationships that are given:

| 1 in | 2.54 cm |
| 1 ft | 12 in |

We now begin with the value that we wish to convert the unit of, i.e., 125.4 cm. We write

| 1 in | 2.54 cm |
| 1 ft | 12 in |

125.4 cm

Since there are two relationships in the list above (one relating 1 in to 2.54 cm and the other relating 1 ft to 12 in), our expression will involve two divisions as shown below.

| 1 in | 2.54 cm |
| 1 ft | 12 in |

$$\underline{} \times \underline{} \times 125.4 \text{ cm}$$

The pair of numbers in each relationship will appear as the dividend and divisor of each of the divisions in the model above. To figure out where the values go, we first note the unit of 125.4, i.e., cm. This requires that we place a value with the unit cm as the divisor of the first division on the right. The only value in the listed relationships that has the unit cm, is 2.54. We write

| 1 in | 2.54 cm ✓ |
| 1 ft | 12 in |

$$\underline{} \times \frac{1 \text{ in}}{2.54 \text{ cm}} \times 125.4 \text{ cm}$$

Note that we have checked off the relationship whose values were just used in populating the first division on the right. This reminds us that the values in this relationship should no longer be used.

The values in the last relationship appear as the dividend and divisor of the remaining division with the value whose unit is the same as the unit of the dividend of the previous division appearing as the divisor of the new division. This gives us

$$\begin{array}{ll} 1 \text{ in} & 2.54 \text{ cm} \quad \checkmark \\ 1 \text{ ft} & 12 \text{ in} \quad \checkmark \end{array}$$

$$\frac{1 \text{ ft}}{12 \text{ in}} \times \frac{1 \text{ in}}{2.54 \text{ cm}} \times 125.4 \text{ cm}$$

To see how the expression above models the problem, we begin by converting the rates into unit rates. This yields the following.

$$0.083\,333\ldots \text{ ft/in} \times 0.393\,700\ldots \text{ in/cm} \times 125.4 \text{ cm}$$

Performing the first multiplication on the right yields the following:[11]

$$0.083\,333\ldots \text{ ft/in} \times 49.370\,078\ldots \text{ in}$$

Performing the last multiplication yields the following:[12]

$$4.11 \text{ ft}$$

The full solution without the intervening comments and explanations is given below:

An Algebraic Solution

L length of the beam (ft)

$$L = \frac{1}{12} \times \frac{1}{2.54} \times 125.4$$

$$L = 4.11 \text{ ft}$$

The beam has a length of 4.11 ft.

---

[11]We let the reader justify the move from $0.393\,700\ldots$ in/cm $\times$ 125.4 cm to $49.370\,078\ldots$ in.

[12]We let the reader justify the move from $0.083\,333\ldots$ ft/in$\times 49.370\,078\ldots$ in to 4.11 ft.

Working with Chemical Reactions

Many problems that relate to computations involving chemical reactions are direct proportion problems.[13] The models for such problems are similar to those given earlier for the solution of everyday life problems and problems that involve conversion of units.

Problem

Consider the balanced reaction for the combustion of propane, C_3H_8:

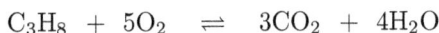

$$C_3H_8 + 5O_2 \rightleftharpoons 3CO_2 + 4H_2O$$

What amount of O_2 was used if the reaction generated 172.7 mol of H_2O?

As shown in Appendix B, this is a direct proportion problem involving the amount of O_2 and the amount of H_2O. Note that, say, doubling the amount of O_2 used, doubles the amount of H_2O produced and that if the reaction generates, say, half as much H_2O, then half as much O_2 was used, etc.

We begin by noting the correspondence between the amount of O_2 and the amount of H_2O as given by the stoichiometric coefficients in the chemical reaction.[14]

$5 \text{ mol}_{O_2} \qquad 4 \text{ mol}_{H_2O}$

Next we begin with the quantity whose value we want to process.

$5 \text{ mol}_{O_2} \qquad 4 \text{ mol}_{H_2O}$
172.7 mol_{H_2O}

Since there is one relationship in the list above (i.e., the relationship between mol_{O_2} and mol_{H_2O}), our expression will have the form of

$$\frac{5 \text{ mol}_{O_2} \qquad 4 \text{ mol}_{H_2O}}{\qquad\qquad} \times 172.7 \text{ mol}_{H_2O}$$

---

[13] For an introduction to chemical reactions please see Appendix B.

[14] This relationship is similar to those that were given in everyday life problems (e.g., the correspondence between 5 pens and $6.00 in the first problem in the subsection on everyday life problems) and conversion of units (e.g., the correspondence between 1 ft and 12 in, in the first problem in the subsection on conversion of units). For problems relating amount of one substance to amount of some other substance in a chemical reaction, such a relationship can always be established using the stoichiometric coefficients in the chemical reaction which may be interpreted in the sense of the one or the mole.

The related values 5 mol_{O_2} and 4 mol_{H_2O} will appear as the dividend and divisor of this division with the one that has the same unit as the unit of 172.7 (i.e., mol_{H_2O}) acting as the divisor, i.e.,

$$5 \ mol_{O_2} \qquad 4 \ mol_{H_2O} \quad \checkmark$$

$$\frac{5 \ mol_{O_2}}{4 \ mol_{H_2O}} \times 172.7 \ mol_{H_2O}$$

To see how this expression models the problem, we begin by converting the rate into a unit rate. This yields the following.

$$1.25 \ mol_{O_2}/mol_{H_2O} \times 172.7 \ mol_{H_2O}$$

The unit rate, 1.25 mol_{O_2}/mol_{H_2O} states that 1.25 mol of O_2 are used for every mol of H_2O generated. Multiplying this unit rate by 172.7 mol_{H_2O} yields the amount of O_2 that is used to generate 172.7 mol_{H_2O}.[15]

The full solution without the intervening remarks is given below.

An Algebraic Solution

n amount of O_2 used (mol)

$$n = \frac{5}{4} \times 172.7$$

$$n = 215.875 \ mol$$

The reaction used 215.875 mol of O_2.

The stoichiometric coefficients in a balanced chemical reaction are also directly proportional to any heat term that may be present.[16]

Problem

Consider the balanced reaction

$$C_3H_8 + 5O_2 \ \rightleftharpoons \ 3CO_2 + 4H_2O + 2220 \ kJ$$

where the stoichiometric coefficients are in moles. The reaction generated 15.8 mol of CO_2. How much heat was generated?

---

[15]The logic behind this move is identical to the logic behind the move from \$2.50/pen × 2 pens to \$5: To find the cost of the pens we multiply the cost of a pen by the number of pens. We let the reader explain the similarity between this logic and the one used in the move from 1.25 $mol_{O_2}/mol_{H_2O} \times$ 172.7 mol_{H_2O} to 215.875 mol_{O_2}.

[16]For an introduction to chemical reactions please see Appendix B.

As shown in Appendix B, this is a direct proportion problem involving the amount of CO_2 and heat generated. Note that the generation of twice as much CO_2 means the generation of twice as much heat and that if the reaction generated one third of the heat then one third as much CO_2 was generated, etc. The technique introduced for modelling direct proportion problems can, therefore, be used in modelling this problem.

We begin by noting the correspondence between the amount of CO_2 and heat generated as given by the balanced reaction.

3 mol_{CO_2} 2220 kJ

Next we begin with the quantity whose value we want to process.

3 mol_{CO_2} 2220 kJ
15.8 mol_{CO_2}

Since there is one relationship (i.e., the relationship between amount of CO_2 and heat generated), our expression will have the form of

3 mol_{CO_2} 2220 kJ
$\underline{\hspace{1.2cm}} \times 15.8 \text{ mol}_{CO_2}$

The related values 3 mol_{CO_2} and 2220 kJ will appear as the dividend and divisor of this division with the one that has the same unit as the unit of 15.8 (i.e., mol_{CO_2}) acting as the divisor, i.e.,

3 mol_{CO_2} $2220 \text{ kJ} \quad \checkmark$

$$\frac{2220 \text{ kJ}}{3 \text{ mol}_{CO_2}} \times 15.8 \text{ mol}_{CO_2}$$

We let the reader justify the setup above. The full solution without the intervening remarks is given below.

An Algebraic Solution

E heat generated (kJ)
$$E = \frac{2220}{3} \times 15.8$$
$$E = 11\,692 \text{ kJ}$$

The reaction generated $11\,692$ kJ of heat.

The stoichiometric coefficients in a chemical reaction relate the amount of one substance to the amount of another substance. In practical applications, however, it is more common to work with the mass or volume of the various substances involved in that reaction. Therefore, if the mass of a certain substance is given, it will have to be converted to the amount of that substance before it can be related to the amount of some other substance by the stoichiometric coefficients in the chemical reaction. Furthermore, if the mass of a substance is sought, any amounts that are calculated using the stoichiometric coefficients must be converted to mass. Conversion of amount to or from mass can be done using the molecular mass or molar mass of the substances involved.

Problem

Consider the balanced reaction

$$C_3H_8 \ + \ 5O_2 \ \rightleftharpoons \ 3CO_2 \ + \ 4H_2O$$

The reaction used 115 g of O_2. What amount of CO_2 was generated? The molar mass of O_2 is 32.00 g/mol.

The various quantities involved (mass of O_2, amount of O_2 and amount of CO_2) are related through direct proportion (twice the mass of O_2 corresponds to twice the amount of O_2 which generates twice the amount of CO_2, etc.) so that the modelling technique given for the solution of direct proportion problems applies.

We begin by noting the correspondences that are given.

$$1 \ mol_{O_2} \qquad 32.00 \ g_{O_2}$$
$$5 \ mol_{O_2} \qquad 3 \ mol_{CO_2}$$

The first relationship comes form the given molar mass of O_2, i.e., 32.00 g/mol. Since *per* mol means *for* 1 mol, the molar mass 32.00 g/mol implies that 1 mol of O_2 has a mass of 32.00 g. The second relationship comes from the stoichiometric coefficients in the balanced reaction which relate the amount of O_2 to the amount of CO_2.

We now begin with the quantity whose value we want to process.

$$1 \ mol_{O_2} \qquad 32.00 \ g_{O_2}$$
$$5 \ mol_{O_2} \qquad 3 \ mol_{CO_2}$$
$$115 \ g_{O_2}$$

Since there are two relationships (i.e., the relationship between the mass of O_2 and the amount of O_2 and the relationship between amount of O_2 and amount of CO_2), our expression will have the form of

$$\frac{1 \ mol_{O_2}}{5 \ mol_{O_2}} \times \frac{32.00 \ g_{O_2}}{3 \ mol_{CO_2}} \times 115 \ g_{O_2}$$

The pair of numbers in each relationship will appear as the dividend and divisor of each of the divisions in the model above. To figure out where the values go, we first note the unit of 115, i.e., g_{O_2}. This requires that we place a value with the unit g_{O_2} as the divisor of the first division on the right. The only value in the listed relationships that has the unit g_{O_2}, is 32.00. We write

$$\begin{array}{ll} 1\ mol_{O_2} & 32.00\ g_{O_2} \quad \checkmark \\ 5\ mol_{O_2} & 3\ mol_{CO_2} \end{array}$$

$$\frac{}{} \times \frac{1\ mol_{O_2}}{32.00\ g_{O_2}} \times 115\ g_{O_2}$$

Note that we have checked off the relationship whose values were just used in populating the first division on the right. This reminds us that the values in this relationship should no longer be used.

The values in the last relationship appear as the dividend and divisor of the remaining division with the value whose unit is the same as the unit of the dividend of the previous division appearing as the divisor of the new division. This gives us

$$\begin{array}{ll} 1\ mol_{O_2} & 32.00\ g_{O_2} \quad \checkmark \\ 5\ mol_{O_2} & 3\ mol_{CO_2} \quad \checkmark \end{array}$$

$$\frac{3\ mol_{CO_2}}{5\ mol_{O_2}} \times \frac{1\ mol_{O_2}}{32.00\ g_{O_2}} \times 115\ g_{O_2}$$

To see how the expression above models the problem, we begin by converting the rates into unit rates. This yields the following.

$$0.6\ mol_{CO_2}/mol_{\ O_2} \times 0.031\,25\ mol_{O_2}/g_{O_2} \times 115\ g_{O_2}$$

Performing the first multiplication on the right yields the following:[17]

$$0.6\ mol_{CO_2}/\ mol_{O_2} \times 3.593\,75\ mol_{O_2}$$

Note that this step converts the given mass of O_2 to the amount of O_2. The need for this conversion is that the stoichiometric coefficients in a chemical reaction relate amounts of substances, not their masses. Therefore, we need to convert the mass of O_2 to amount of O_2 before we can use the stoichiometric coefficients to work out the corresponding amount of CO_2.

Performing the last multiplication yields the following:[18]

$$2.156\,25\ mol_{CO_2}$$

The full solution without the intervening remarks is given below.

---

[17] We let the reader explain the logic behind this move.
[18] We let the reader explain the logic behind this move.

An Algebraic Solution

n amount of CO_2 generated (mol)

$$n = \frac{3}{5} \times \frac{1}{32.00} \times 115$$

$n = 2.156\,25$ mol

The reaction generated $2.156\,25$ mol of CO_2.

We present one more example.

Problem

Consider the balanced reaction

$$C_3H_8 + 5O_2 \rightleftharpoons 3CO_2 + 4H_2O$$

The reaction generated 42.5 g of H_2O. What mass of O_2 was used? The molar masses of H_2O and O_2 are 18.016 g/mol and 32.00 g/mol respectively.

An Algebraic Solution

m mass of O_2 used (g)

$$m = \frac{32.00}{1} \times \frac{5}{4} \times \frac{1}{18.016} \times 42.5$$

$m = 94.36$ g

The reaction used 94.36 g of O_2.

We let the reader justify the logic behind the model in the solution above.

Fractional Problems

The discussion in this section relates to problems that require that we compute a fraction, decimal fraction or percentage of the value of a quantity. Examples of such problems are finding $\frac{2}{3}$ of the number of students in the class, 0.25 of the volume of a solution and 15% of the price of a hat. The key to working with such problems is the understanding that they can be worked out by multiplying the given fraction, decimal fraction or percentage by the value of the quantity whose piece we wish to compute. As an example, if there are 36 students in the class, then to find $\frac{2}{3}$ of the number of students in the class, we can multiply $\frac{2}{3}$ by 36, i.e., we can model the problem

There are 36 students in the class. Calculate $\frac{2}{3}$ of the number of students in the class.

as[19]

n number of students sought (1)

$$n = \frac{2}{3} \times 36$$

We can explain the correspondence between the English phrase and its corresponding mathematical model by appealing to the semantics of the word *of* when it appears in such constructs. Note that when we speak of 2 *of something* we mean $2 \times$ *the value of the thing* and when we speak of 3 *of something* we mean $3 \times$ *the value of the thing*, and so on. Extending this interpretation of the word *of* to contexts that involve fractions, decimal fractions or percentages leads to the equivalence between $\frac{2}{3}$ of something and $\frac{2}{3} \times$ *the value of the thing*, 0.25 of something as $0.25 \times$ *the value of the thing* and 15% *of something* as $15\% \times$ *the value of the thing*, and so on.

To see how multiplication captures the semantics of *fraction of, decimal fraction of* and *percentage of*, we note that when we speak of $\frac{2}{3}$ of the value of a quantity, we mean to imply that the value of the quantity should be divided into 3 parts and that 2 parts should be taken. Therefore, to solve the problem above, we need to divide the number of students, i.e., 36, into 3 equal parts, which leads to $36 \div 3$ or 12 students per part, and then take two parts, i.e., 2×12, to arrive at 24 students as the answer to the problem. Note that, as shown below, multiplication of $\frac{2}{3}$ by 36 takes both steps: Reduction of 36 and 3 performs the first step, i.e., $36 \div 3$, and subsequent multiplication of 12 by 2 performs the second step, i.e., 2×12.

$$\frac{2}{\underset{1}{3}} \times \overset{12}{\cancel{36}} = 24$$

Similar arguments can be given for the equivalence between the phrases *a decimal fraction of* and *a percentage of* and their corresponding mathematical counterparts *decimal fraction*$\times$ and *percentage*$\times$. See the exercise set that follows for detail.

---

[19]This has led some to claim that the word *of* in a word problem translates to $\times$ in the corresponding mathematical model. It would be more accurate to say that English phrases such as *a fraction of, a decimal fraction of* and *a percentage of* in a word problem translate to *fraction*$\times$, *decimal fraction*$\times$ or *percentage*$\times$ in the corresponding mathematical model.

We are now ready to discuss the modelling of percent problems. We will begin with percent problems that involve money and then extend the discussion to general percent problems.

Percent Problems involving Discount and Tax

Consider the following problem.

Problem

A hat, priced at \$24.50, is on sale at 10% off. Calculate the amount of discount.

The relevant quantities involved in such a problem are the price, the rate of discount and the amount of discount. The simplest way to relate these quantities is to argue that *the amount of discount is some percentage of the price*. This translates verbatim into an equation:

$$\underbrace{\text{amount of discount}}\text{ is }\underbrace{\text{some percentage}}\text{ of }\underbrace{\text{the price}}$$

$$a \qquad = \qquad 10\% \qquad \times \qquad 24.50$$

where a represents the amount of discount in dollars.

Note the manner in which elements of the language map onto mathematical symbols that express the same idea. Two languages for expressing the same thought with the second being much more efficient at doing so.

It is common for us to write the model using decimals in place of percentages. Adoption of this convention leads to the following model for the problem above:[20]

$$a = 0.1 \times 24.50$$

The full solution without the intervening explanations in given below.

---

[20]Note that ten percent of the price is the same as one tenth of the price.

An Algebraic Solution

> a amount of discount (\$)
> $a = 0.1 \times 24.50$
>
> $a = \$2.45$
>
> The amount of discount is \$4.25.

Problem

> A hat, priced at \$24.50, is on sale at 10% off. Calculate the sale price.

Since the hat is on sale at 10% of the price, we will end up paying 90% of the price. This should make sense. In fact just as we can say

> amount of discount is 10% of the price

we can claim that

> sale price is 90% of the price

As the reader might have suspected, the 90% in the claim above results from the subtraction of 10% from 100% (with 100% representing the price of the hat). Note that if we cut 10% of the price of the hat, we will be left with 90% of the price of the hat which is the amount that we will pay, i.e., sale price.

The discussion above leads to the following model.

> $p_s = (100\% - 10\%)(24.50)$

where p_s represents the sale price in dollars.

Note that we have written $(100\% - 10\%)$ in place of 90%. This we do for two reasons: First, one should explain where 90% comes from, whether or not one thinks the answer is obvious. More important, later we will come across problems where the price and sale price are both given and one needs to compute the rate of discount. In such a case our model above will have the factor $(100\% - r)$ with r representing the rate of discount. It is sensible to get used to constructing such a factor by being explicit about it now.[21]

The full solution using decimals in place of percentages is given below.

---

[21] The relationship given above in the body of the text is more efficient than the one that states that *the sale price is the difference between the price and the amount of discount*

An Algebraic Solution

p_s sale price ($)
$p_s = (1 - 0.1)(24.50)$
$p_s = 0.9 \times 24.50$
$p_s = \$22.05$
The sale price is \$22.05.

Our next example is one that relates to the amount of tax.

Problem

A hat is priced at \$24.50. Tax is added at 13% of the price. Calculate the amount of tax.

The relevant quantities involved in such a problem are the price, the rate of tax and the amount of tax. The simplest way to relate these quantities is to argue that *the amount of tax is some percentage of the price*. This translates verbatim into an equation:

$a = 13\% \times 24.50$

where a represents the amount of tax in dollars. We let the reader justify the model. The full solution using decimals in place of percentages is given below.

An Algebraic Solution

a amount of tax ($)
$a = 0.13 \times 24.50$

$a = \$3.19$

The amount of tax is \$3.19.

---

with the amount of discount expressed as 10% *of the price*, i.e.,

$p_s = 24.50 - 10\% \times 24.50$

For one thing, the repetition of 24.50 is unwelcome as, in a problem where the price is unknown, the unknown appears multiple times. Furthermore, the model leaves a strong impression that the relationship between the price and sale price is one of superposition while the formulation given in the body of the text makes it clear that the relationship is one of direct proportion. And, as we will see later, such constructs become unwieldy when the solution to the problem consists of a sequence of two or more proportions. See our comments later where we discuss problems in which discount is followed by tax.

Problem

A hat is priced at $24.50. Tax is added at 13% of the price. Calculate the grand total.

Since tax is added at 13 of the price%, we will end up paying 113% of the price. This should make sense. In fact just as we can say

amount of tax is 13% of the price

we can claim that

grand total is 113% of the price

As the reader might have suspected, the 113% in the claim above results from the addition of 13% to 100% (with 100% representing the price of the hat). Note that if we add 13% of the price of the hat, we will end up with 113% of the price of the hat as the grand total.

The discussion above leads to the following model.

$$t = (100\% + 13\%)(24.50)$$

where t represents the grand total in dollars.

Note that we have written $(100\% + 13\%)$ in place of 113%. This we do for two reasons: First, one should explain where 113% comes from, whether or not one thinks the answer is obvious. More important, later we will come across problems where the price and the grand total are both given and one needs to compute the rate of tax. In such a case our model above will have the factor $(100\% + r)$ with r representing the rate of tax. It is sensible to get used to constructing such a factor by being explicit about it now.[22]

The full solution using decimals in place of percentages is given below.

---

[22]The relationship given above in the body of the text is more efficient than the one that states that *the grand total is the sum of the price and the amount of tax* with the amount of tax expressed as 13% *of the price*, i.e.,

$$t = 24.50 + 13\% \times 24.50$$

For one thing, the repetition of 24.50 is unwelcome as, in a problem where the price is unknown, the unknown appears multiple times. Furthermore, the model leaves a strong impression that the relationship between the price and the grand total is one of superposition while the formulation given in the body of the text makes it clear that the relationship is one of direct proportion. And, as we will see later, such constructs become unwieldy when the solution to the problem consists of a sequence of two or more proportions. See our comments later where we discuss problems in which discount is followed by tax.

An Algebraic Solution

t grand total ($)
$$t = (1 + 0.13)(24.50)$$
$$t = 1.13 \times 24.50$$
$$t = \$27.69$$

The grand total is $27.69.

Here is a problem involving discount followed by tax.

Problem

A hat, priced at $24.50, is on sale at 10% off. Tax is added at 13% of the sale price. Calculate the amount of tax.

Note that the problem states that the amount of tax is 13% of the *sale* price. Since the sale price is 90% of the price, we can model the problem as

$$a = 13\% (100\% - 10\%)(24.50)$$

The first factor on the right side of the expression on the right side of the model above, i.e., 24.50, represents the price of the hat. This is shown below.

$$13\% \quad \times \quad (100\% - 10\%) \quad \times \quad \underbrace{(24.50)}_{\text{Price}}$$

Moving one factor to the left, the subexpression $(100\% - 10\%)(24.50)$ simplifies to 90% of the price, i.e., the sale price. This is shown below.

$$13\% \quad \times \quad \underbrace{\overbrace{(100\% - 10\%)}^{90\%} \quad \times \quad \overbrace{(24.50)}^{\text{the price}}}_{\text{Sale Price}}$$

We now move another factor to the left to get the full expression $13\% (100\% - 10\%)(24.50)$. This represents 13% of the sale price, i.e., the amount of tax. This is shown below.

$$\underbrace{\overbrace{13\%}^{13\%} \quad \times \quad \overbrace{(100\% - 10\%) \quad \times \quad (24.50)}^{\text{the sale price}}}_{\text{Amount of Tax}}$$

The full solution without the intervening comments using decimals in place of percentages is given below.

An Algebraic Solution

a amount of tax (\$)
$a = 0.13\,(1 - 0.1)\,(24.50)$

$a = 0.13 \times 0.9 \times 24.50$
$a = \$2.87$

The amount of tax is \$2.87.

Problem

A hat, priced at \$24.50, is on sale at 10% off. Tax is added at 13% of the sale price. Calculate the grand total.

An Algebraic Solution

t grand total (\$)
$t = (1 + 0.13)\,(1 - 0.1)\,(24.50)$

$t = 1.13 \times 0.9 \times 24.50$
$t = \$24.92$

The grand total is \$24.92.

We let the reader make sense of the model above.

General Percent Problems

Percent problems often require that we either work out the size of a piece, find the amount that is left over after we cut off a piece, or compute the total that one arrives at after adding a piece. The first type is similar to money problems in which we seek to find the amount of discount or the amount of tax. The second type is similar to money problems in which we seek to find the sale price. The third type is similar to money problems in which we seek to find the grand total.

Problem

In a sample of 240 frogs, 10% have green skin. How many frogs in the sample have green skin?

The wording of this problem implies the following: The number of green-skinned frogs in the sample is 10% of the total number of frogs. We translate this verbatim into a mathematical equation:

$$n = 10\% \times 240$$

where n represents the number of green-skinned frogs in the sample. The full solution without the intermediate comments with decimals in place of percentages is given below.

An Algebraic Solution

n number of green-skinned frogs in the sample (1)
$n = 0.1 \times 240$

$n = 24$

There are 24 green-skinned frogs in the sample.

Note that this problem is similar to money problems in which we seek to find the amount of discount or the amount of tax. Indeed, the model above would work for the problem *A hat, priced at $240.00, is on sale at 10% off. Calculate the amount of discount.* The following equation models the problem:

$$a = 0.1 \times 240$$

where a represents the amount of discount with unit dollar.

Problem

In a sample of 240 frogs, 10% have green skin. How many frogs in the sample do *not* have green skin?

One can argue that, since 10% of the frogs in the sample have green skin, then 90% of the frogs in the sample do *not* have green skin. We traslate this verbatim into a mathematical equation:

$$n = (100\% - 10\%)(240)$$

where n represents the number of *non*-green-skinned frogs in the sample. This leads to the following algebraic solution.

An Algebraic Solution

n number of non-green-skinned frogs in the sample (1)
$n = (1 - 0.1)(240)$

$n = 0.9 \times 240$
$n = 216$

There are 216 non-green-skinned frogs in the sample.

Note that this problem is similar to money problems in which we seek to find the sale price. Indeed, the model above would work for the problem *A hat, priced at \$240, is on sale at 10% off. Calculate the sale price.* The following equation models the problem:

$p_s = (1 - 0.1)(240)$

where p_s represents the sale price with unit dollar.

Problem

The population of a town grew by 13% over the past decade. How many more people live in the town now compared to ten years ago? The town had a population of 120 000 ten years ago.

The word problem above implies that *the amount of increase over the last decade was 13% of the population ten years ago.* This translates verbatim into a mathematical equation.

$n = 13\% \times 120\,000$

where n represents the amount of increase. We let the reader explain the correspondence between the sentence *the amount of increase over the last decade was* 13% *of the population ten years ago* and its mathematical counterpart $n = 13\% \times 120\,000$. Here is an algebraic solution:

An Algebraic Solution

n amount of increase in population of town (1)
$n = 0.13 \times 120\,000$

$n = 15\,600$

The population of the town increased by 15 600 over the last ten years.

Note that this problem is similar to money problems in which we seek to find the amount of tax. Indeed, the model above would work for the problem *An ancient hat is priced at $120 000.*[23] *Tax is added at 13% of the price. Calculate the amount of tax.* The following equation models the problem:

$$a = 0.13 \times 120\,000$$

where a represents the amount of tax with unit dollar.

Problem

The population of a town grew by 13% over the past decade. What is the population of the town now if it had a population of 120 000 ten years ago?

Such wordings imply that *the population today is 113% of the population ten years ago.* This translates verbatim into a mathematical equation.

$$n = (100\% + 13\%)(120\,000)$$

where n represents the population of the town today. Here is the full solution.

An Algebraic Solution

n population of the town today (1)
$$n = (1 + 0.13)(120\,000)$$
$$n = 1.13 \times 120\,000$$
$$n = 135\,600$$

The town has a population of 135 600 today.

Note that this problem is similar to money problems in which we seek to find the grand total. Indeed, the model above would work for the problem *An ancient hat is priced at $120 000. Tax is added at 13% of the price. Calculate the grand total.* The following equation models the problem:

$$t = (1 + 0.13)(120\,000)$$

where t represents the grand total with unit dollar.

As with money problems, multi-step problems are common. Here is an example.

---

[23] :)

Problem

In a sample of 160 frogs, 10% have green skin. Of those that have green skin, 25% have red eyes. How many red-eyed, green-skinned frogs are there in the sample?

Note that the problem states that the number of red-eyed, green-skinned frogs in the sample is equal to 25% of the number of green-skinned frogs in the sample. Since the number of green-skinned frogs in the sample is 10% of the total number of frogs in the sample, we can model the problem as

$$n = 25\% \times 10\% \times 160$$

The first factor on the right side of the expression on the right side of the model above, i.e., 160, represents the number of frogs in the sample. This is shown below.

$$25\% \qquad \times \qquad 10\% \qquad \times \qquad \underbrace{160}_{\text{Total Number of Frogs}}$$

Moving one factor to the left, the subexpression $10\% \times 160$ represents the number of green-skinned frogs in the sample. This is shown below.

$$25\% \qquad \times \qquad \underbrace{\overbrace{10\%}^{10\%} \qquad \times \qquad \overbrace{160}^{\text{total number of frogs}}}_{\text{Number of Green-Skinned Frogs}} \quad \overset{\text{of}}{}$$

Moving another factor to the left, the subexpression $25\% \times 10\% \times 160$ represents 25% of the number of green-skinned frogs, i.e., the number of red-eyed, green-skinned frogs. This is shown below.

$$\underbrace{\overbrace{25\%}^{25\%} \qquad \times \qquad \overbrace{10\% \qquad \times \qquad 160}^{\text{the number of green-skinned frogs}}}_{\text{Number of Red-Eyed, Green-Skinned Frogs}} \quad \overset{\text{of}}{}$$

The full solution without the intervening comments using decimals in place of percentages is given below.

An Algebraic Solution

n number of red-eyed, green-skinned frogs in sample (1)
$n = 0.25 \times 0.1 \times 160$

$n = 4$

The are 4 red-eyed, green-skinned frogs in the sample.

Problem

In a sample of 160 frogs, 10% have green skin. Of those that have green skin, 25% have red eyes. How many green-skinned frogs in the sample do not have red eyes?

An Algebraic Solution

n number of non-red-eyed, green-skinned frogs in sample (1)
$n = (1 - 0.25)(0.1)(160)$

$n = 0.75 \times 0.1 \times 160$
$n = 12$

The are 12 non-red-eyed, green-skinned frogs in the sample.

We let the reader justify the model above.

Problem

The population of a town increased by 10.3% in the seventies and then decreased by 5.2% in the eighties. What is the population of the town at the end of the eighties if the town had a population of 310 000 at the beginning of the seventies?

An Algebraic Solution

n population of the town at the end of eighties (1)
$n = (1 - 0.052)(1 + 0.103)(310\,000)$

$n = 0.948 \times 1.103 \times 310\,000$
$n = 324\,150$

At the end of the eighties, the town had a population of 324 150.

We let the reader justify the model above.

Multiplication of Rates

We now turn our attention to problems whose solutions lie in working out certain subexpressions of the model for direct proportion problems.

The reader might have noticed that the models discussed so far for the solution of the types of problems we just covered have the form of

$$f_n \cdots f_2 f_1 v$$

where v represents the value of the quantity that is being processed and the factors represented by the symbol f_i, $i = 1, 2, \ldots, n$, represent a sequence of factors that v is multiplied by. The key to the solution of a class of problems involving direct proportion lies in the value of those subexpressions of $f_n \cdots f_2 f_1 v$ that consist of a contiguous sequence of the f_i.[24] Here is an example of such a problem.

Problem

Find a conversion factor for conversion of the unit of length from feet to centimetres given that

$$1 \text{ ft} = 12 \text{ in}$$
$$1 \text{ cm} = 0.394 \text{ in}$$

The related values listed above can be used to convert the unit of length from feet to centimetres and vice versa. As an example, with l_{ft} representing the length in feet and l_{cm} representing the length in centimetres, the expression

$$l_{cm} = \frac{1}{0.394} \times \frac{12}{1} \times l_{ft}$$

converts the unit of length from feet to centimetres: Multiplication of the length in feet, i.e., l_{ft}, by $\frac{12}{1}$ converts the unit of length from feet to inches and multiplication of the result by $\frac{1}{0.394}$ converts the unit of length from inches to centimetres. This is shown below:

$$\xrightarrow{\times \frac{12}{1}} \qquad \xrightarrow{\times \frac{1}{0.394}}$$

length (ft) $\xrightarrow{\hspace{2cm}}$ length (in) $\xrightarrow{\hspace{2cm}}$ length (cm)

The problem above requires that we find a conversion factor that directly relates the length in feet to the length in centimetres. This is shown below:

---

[24]By a contiguous sequence of f_i we mean a sequence such as $f_4 f_3$ and $f_3 f_2 f_1$ where there are no gaps in the indices between the largest index and the smallest index.

$$\text{length (ft)} \xrightarrow{\times \frac{?}{?}} \text{length (cm)}$$

The solution to this problem lies in the simplification of the expression on the right hand side of

$$l_{cm} = \frac{1}{0.394} \times \frac{12}{1} \times l_{ft}$$

This leads to

$$l_{cm} = 30.5 \times l_{ft}$$

The factor 30.5 has a unit of cm/ft.[25] As such, it can be used to directly convert the unit of length from feet to centimetres. Incidentally, this also tells us that 1 ft = 30.5 cm as is clear from the reading of "/" as *per*. The full solution to the problem above without the intervening comments is given below:

An Algebraic Solution

> r conversion factor from feet to centimetres (cm/ft)
> $$r = \frac{1}{0.394} \times \frac{12}{1}$$
> $$r = 30.5 \text{ cm/ft}$$
> The conversion factor from feet to centimetres is 30.5 cm/ft.

As a second example of a problem whose solution requires that we multiply the factors represented by f_i in $f_n \cdots f_2 f_1 v$, we present the following problem:

Problem

> 75% of the eligible voters voted. Of those who did not vote, 80% cited long voting queues as the reason for not voting. What percentage of eligible voters did not vote, citing long voting queues as the reason for not voting?

The following equation models the problem:

$$r = 0.8 \, (1 - 0.75)$$

---

[25] We ask the reader to justify this claim.

where r represents the percentage of eligible voters who did not vote and cited long voting queues as the reason for not voting.

Let us take a closer look at the model above. Following the statement of the problem, we begin with the number of eligible voters. Since this is not known, we will use the quantity symbol n_e to represent it with. We write

$$n_e$$

The word problem states that 75% of the eligible voters voted. Since we are interested in those who did *not* vote, we write

$$(100\% - 75\%)\, n_e$$

and we can compute the number of eligible voters who did not vote and cited long voting queues as the reason for not voting using the expression

$$80\% \, (100\% - 75\%)\, n_e$$

Using the symbol n to represent the number of eligible voters who did not vote and cited long voting queues as the reason for not voting, we have

$$n = 80\% \, (100\% - 75\%)\, n_e$$

In the expression on the right side of the model above, multiplication of n_e by the factor $(100\% - 75\%)$ yields the number of eligible voters who did not vote.[26] Following this, multiplication of the number of eligible voters who did not vote by 80% yields the number of eligible voters who did not vote and cited long voting queues as the reason for not voting. The problem above requires that we find a percentage that directly relates the number of eligible voters to the number of eligible voters who did not vote and cited long voting queues as the reason for not voting. The solution to this problem lies in the simplification of the expression on the right hand side of

$$n = 80\% \, (100\% - 75\%)\, n_e$$

This leads to

$$n = 80\% \times (100\% - 75\%) \times n_e$$
$$n = 0.8 \times (1 - 0.75) \times n_e$$
$$n = 0.8 \times 0.25 \times n_e$$
$$n = 0.2 \times n_e$$
$$n = 20\% \times n_e$$

The last line above reads as follows: *The number of eligible voters who did not vote and cited long voting queues as the reason for not voting is 20% of the number of eligible voters.* This provides us with the answer that we have been seeking. The full solution to the problem without the intervening comments is given below.

---

[26]The use of $100\% - 75\%$, which simplifies to 25%, reflects the fact that, if 75% of the eligible voters voted, then $100\% - 75\%$ or 25% of them did not vote.

An Algebraic Solution

r percentage of eligible voters who did not vote and cited long
    voting queues as the reason for not voting (1)

$r = 0.8 \,(1 - 0.75)$
$r = 0.8 \times 0.25$
$r = 0.2$

20% of eligible voters did not vote and cited long voting queues
as the reason for not voting.

The example above illustrates the circumstances under which it is appropriate to multiply percentages: Given the values of three quantities, v_A, v_B and v_C, if v_A is a certain percentage of v_B and v_B is a certain percentage of v_C, then, to find out what percentage of v_A the value v_C is, we can multiply the percentages that relate v_A to v_B, and v_B to v_C. As an example, if v_B is 20% of v_A and v_C is 10% of v_B, then v_C is 10%×20% or 2% of v_A.[27] This is shown below:

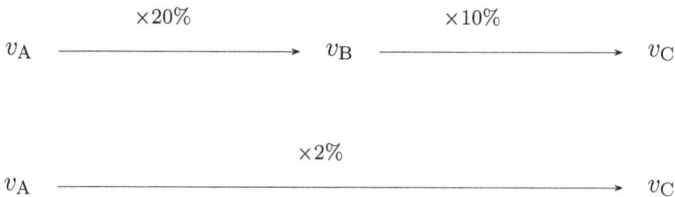

Subexpressions

As stated earlier, in a given expression, the value of the quantity represented by a term is directly proportional to the values of the quantities that are represented by the factors within it. As an example, the expression

$$q_4 q_3 q_2 q_1$$

is directly proportional to q_1, q_2, q_3 and q_4. Further to this, we can show that the value of the quantity represented by a term is directly proportional to the values of the quantities that are represented by any combination of its factors. As an example, in addition to being directly proportional to q_1, q_2, q_3 and q_4, the expression $q_4 q_3 q_2 q_1$ is directly proportional to $q_2 q_1$, $q_3 q_2$, $q_3 q_2 q_1$, $q_4 q_2 q_1$, etc., and of course the expression $q_4 q_3 q_2 q_1$ itself. As an example of how this is the case, consider the subexpression $q_3 q_2$. We can show that $q_4 q_3 q_2 q_1$ is

---

[27]Note that 20%×10% simplifies to 0.2×0.1 or 0.02 which corresponds to a value of 2%.

directly proportional to q_3q_2 by placing the subexpression within brackets which effectively turns the combo into a single factor. This is shown below.

$$q = q_4 (q_3q_2) q_1$$

The expression on the right side now breaks into three factors: q_4, q_3q_2, and q_1. It should now be apparent that $q_4q_3q_2q_1$ is directly proportional to (q_3q_2) since the subexpression is being multiplied by. Indeed, if (q_3q_2) increases or decreases in size by a factor of, say, 2, then we will be multiplying by a value that is 2 times larger, resulting in a value for $q_4q_3q_2q_1$ that is 2 times larger and so on.

At times we may choose to name the quantity whose value is represented by an expression. As an example, we may choose to refer to the expression $q_4q_3q_2q_1$ as q. This leads to the following equation:

$$q = q_4q_3q_2q_1$$

Since q is the same as $q_4q_3q_2q_1$, then any conclusions that we make about $q_4q_3q_2q_1$ will hold true for q as well. As an example, in the equation above, we can say that the expression $q_4q_3q_2q_1$ is directly proportional to the subexpressions q_1, q_2, q_4q_3, and so on. Since q is another way to refer to $q_4q_3q_2q_1$, then we can say that q is directly proportional to q_1, q_2, q_4q_3, and so on.

When a term involves a division, we may choose to further analyze the factor into factors in its dividend and factors in its divisor. As an example, when we analyze the expression

$$\frac{q_3q_2q_1}{q_4q_5q_6}$$

we see the expression as a single term that contains a single factor. This factor may, however, be further analyzed into factors in its dividend, i.e., q_1, q_2, and q_3, and factors in its divisor, i.e., q_4, q_5, and q_6. It can now be shown that $\frac{q_3q_2q_1}{q_4q_5q_6}$ is directly proportional to any of its subexpressions that consists of one factor (e.g., q_3, $\frac{1}{q_6}$, etc.) two factors (e.g., q_3q_1, $\frac{q_2}{q_6}$, etc.), three factors (e.g., $q_3q_2q_1$, $\frac{q_1}{q_4q_6}$, etc.), and so on, including the full expression itself, i.e., $\frac{q_3q_2q_1}{q_4q_5q_6}$. See the exercises for a proof.

As before, we may choose to name the quantity whose value is represented by the full expression. Referring to the expression $\frac{q_3q_2q_1}{q_4q_5q_6}$ as q allows us to write

$$q = \frac{q_3q_2q_1}{q_4q_5q_6}$$

and to conclude that any conclusions that hold true about $\frac{q_3q_2q_1}{q_4q_5q_6}$ will hold true about q as well.

Finally, note that direct proportion is symmetric. If the value of one quantity is directly proportional to the value of another quantity, then the value of the second quantity is directly proportional to the value of the first quantity. We will justify this claim in the pages ahead.

Exercise Set 7

1. Justify the equivalence between the phrase 0.25 *of* 20 and the mathematical expression 0.25 × 20 as well as the equivalence between the phrase 25% *of* 20 and the mathematical expression 25% × 20.

2. Solve each of the following problems using the algebraic technique.

   a. 4.5 L of a solution has a mass of 12.8 g. At this rate, what is the mass of 15.6 L of the solution?

   b. 4.5 L of a solution has a mass of 12.8 g. At this rate, what is the volume of 62.8 g of the solution?

   c. The recipe calls for 2.5 tablespoons of oil for 1.5 cups of pop corn kernels. At this rate, how much oil should be used for 3.5 cups of pop corn kernels?

   d. The recipe calls for 2.5 tablespoons of oil for 1.5 cups of pop corn kernels. At this rate, how many cups of pop corn kernels should be used if we have 1.75 tablespoons of oil?

   e. You used 7.5 L of fuel to cover a distance of 80 km. At this rate, what volume of fuel will be needed to cover the remaining distance of 135 km?

   f. You used 7.5 L of fuel to cover a distance of 80 km. At this rate, how far can you drive if you have 4.5 L of fuel left?

   g. Stephanie made $231 for 16.5 h of work. At this rate, how much more can she make if she works an extra 3.5 h?

   h. Stephanie made $231 for 16.5 h of work. At this rate, how many hours of extra work will she have to log in to make an extra $69.00?

   i. You have 15.75 L of fuel left. At the speed that you are driving, your car uses 6.85 L of fuel to cover 100 km. How long does it take to run out of fuel if you are driving at a speed of 120 km/h?

   j. Your car's rate of consumption of fuel is 7.2 L for 100 km. The cost of fuel is $1.136/L. What is the cost of fuel for a trip to a town 325 km away?

3. Solve each of the following problems using the algebraic technique.

   a. The presentation was 135 min long. Convert this to hours given that

   $$1 \text{ h } = \text{ 60 min}$$

   b. The newborn has a mass of 3.2 kg. Convert this to pounds if

   $$1 \text{ kg } \cong \text{ 2.2 lb}$$

   c. The newborn has a mass of 3.2 kg. Convert this to pounds if

   $$1 \text{ lb } \cong \text{ 0.454 kg}$$

   d. You have 1200 mL of a medication. Convert this to ounces given that

   $$1 \text{ mL } \cong \text{ 0.034 oz}.$$

   e. You have 1200 mL of a solution. Convert this to ounces given that

   $$1 \text{ oz } \cong \text{ 29.6 mL}$$

   f. The scar has a length of 1.5 in. Convert this to mm given that

   $$1 \text{ in } \cong \text{ 2.54 cm}$$
   $$1 \text{ mm } = \text{ 0.1 cm}$$

   g. The liquid has a mass of 3.8 kg. Convert this to lb given that

   $$1 \text{ g } = \text{ 0.001 kg}$$
   $$1 \text{ g } \cong \text{ 0.0022 lb}$$

   h. You have $250 Can. Convert this to Euro assuming exchange rates of

   $$\$1 \text{ Can } = \text{ \$0.97444 US}$$
   $$\$1 \text{ US } = \text{ 0.71629 €}$$

   i. You have $250 Can. Convert this to Euro assuming exchange rates of

   $$\$1 \text{ US } = \text{ \$1.02523 Can}$$
   $$1 \text{ € } = \text{ \$1.39509 US}$$

   j. You measure a pressure of 1.2 atm. Convert this to kPa given that

   $$1 \text{ atm } = \text{ 101325 Pa}$$
   $$1 \text{ Pa } = \text{ 0.001 kPa}$$

k. The distance from New York City to Los Angeles is 2780 mi. Convert this to km if

$$1 \text{ mi} \cong 1760 \text{ yd}$$
$$1 \text{ m} = 1.09 \text{ yd}$$
$$1 \text{ km} = 1000 \text{ m}$$

l. The chemical reaction generated 3500 kJ of heat. Convert this to Calories given that

$$1 \text{ kJ} = 1000 \text{ J}$$
$$1 \text{ cal} = 4.184 \text{ J}$$
$$1 \text{ cal} = 0.001 \text{ Cal}$$

4. Solve each of the following problems using the algebraic technique.

a. Calculate the mass of 4.8 mol of $C_6H_{12}O_6$. The molar mass of $C_6H_{12}O_6$ is 180.156 g/mol.

b. What amount of $C_6H_{12}O_6$ corresponds to 76 g of $C_6H_{12}O_6$? The molar mass of $C_6H_{12}O_6$ is 180.156 g/mol.

c. Calculate the mass of 1900 mol of O_2. The molar mass of O_2 is 32 g/mol.

d. What amount of O_2 corresponds to 1200 g of O_2? The molar mass of O_2 is 32 g/mol.

e. Calculate the mass of 20 molecules of NO_3. The molecular mass of NO_3 is 62.01 amu/molecule.

f. The molecular mass of SO_2 is 64.07 amu/molecule. How many molecules of SO_2 have a mass of 80 087.5 amu?

5. Solve each of the following problems using the algebraic technique.

a. Consider the chemical reaction for the combustion of C_3H_6:

$$2C_3H_6 + 9O_2 \rightleftharpoons 6CO_2 + 6H_2O$$

What amount of CO_2 is generated by the combustion of 4.5 mol of C_3H_6?

b. Aluminum oxide, Al_2O_3, forms according to the reaction

$$4Al + 3O_2 \rightleftharpoons 2Al_2O_3$$

What amount of aluminum was used in the generation of 520 mol of aluminum oxide?

c. H_2SO_4 dissolves in water to generate H^+ and SO_4^{2-} ions according to the equation

$$H_2SO_4 \rightleftharpoons 2H^+ + SO_4^{2-}$$

What amount of H^+ ions is generated if the dissociation generates 120 mol of SO_4^{2-}?

d. Formation of water from H_2 and O_2 follows the reaction

$$2H_2 + O_2 \rightleftharpoons 2H_2O + 518 \text{ kJ}$$

What amount of H_2O was generated if the reaction generated 1420 kJ of heat?

e. Ozone, O_3, can be generated according to the reaction

$$O_2 + O \rightleftharpoons O_3$$

What amount of ozone can be generated from the use of 125 mol of O_2?

f. Consider the chemical reaction

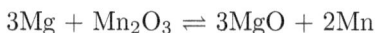

$$3Mg + Mn_2O_3 \rightleftharpoons 3MgO + 2Mn$$

What amount of Mg is needed to generate 42.5 mol of Mn?

g. Consider the reaction of nitrogen and oxygen to generate nitrogen oxide:

$$N_2 + O_2 + 44 \text{ Cal} \rightleftharpoons 2NO$$

How much heat is taken up by the generation of 32.8 mol of NO?

6. Solve each of the following problems using the algebraic technique.

a. The chemical reaction for the combustion of ethane, C_2H_6, is

$$2C_2H_6 + 5O_2 \rightleftharpoons 4CO_2 + 2H_2O + 1888 \text{ kJ}$$

   i. What amount of CO_2 is generated by the combustion of 82.7 g of C_2H_6? The molar mass of C_2H_6 is 30.068 g/mol.

   ii. What mass of C_2H_6 is used if the reaction uses 122 mol of O_2? The molar mass of C_2H_6 is 30.068 g/mol.

   iii. What amount of H_2O is generated if the reaction generates 42.8 g of CO_2? The molar mass of CO_2 is 44.01 g/mol.

b. Ethanol, the alcohol in drinking beverages, is metabolized according to the following chemical reaction

$$C_2H_5OH + 3O_2 \rightleftharpoons 2CO_2 + 3H_2O + 1790 \text{ kJ}$$

   i. What mass of CO_2 is generated by the use of 154.7 g of O_2? The molar masses of CO_2 and O_2 are 44.01 g/mol and 32.00 g/mol respectively.

   ii. What mass of H_2O is generated if the reaction uses 1800 g of C_2H_5OH? The molar masses of H_2O and C_2H_5OH are 18.016 g/mol and 46.068 g/mol respectively.

   iii. What mass of O_2 is used if the reaction uses 120 g of C_2H_5OH? The molar mass of C_2H_5OH is 46.068 g/mol.

c. Incomplete combustion of ethyne, C_2H_2, generates carbon monoxide, CO, according to the reaction

$$2C_2H_2 + 3O_2 \rightleftharpoons 4CO + 2H_2O + 2511 \text{ kJ}$$

i. What mass of carbon monoxide is generated if the reaction generates 12 000 kJ of heat? The molar mass of CO is 28.01 g/mol.

ii. How much heat is generated by the use of 450 g of C_2H_2? The molar mass of C_2H_2 is 26.036 g/mol.

d. Water, H_2O, decomposes according to the reaction

$$2H_2O + 518 \text{ kJ} \rightleftharpoons 2H_2 + O_2$$

i. What mass of water is used if the reaction takes up 38 500 kJ of heat? The molar mass of water is 18.016 g/mol.

ii. How much heat is taken up by the decomposition process if the reaction generates 72.6 g of H_2? The molar mass of H_2 is 2.016 g/mol.

7. Solve the following problems using the algebraic technique.

a. A hat, priced at $32.99, is on sale at 15% off. Calculate the amount of discount.

b. A chair, priced at $125.00 is on sale at 10% off. Calculate the amount of discount.

c. A soccer ball is priced at $25.50. It is on sale at 10% off. Calculate the sale price.

d. A book, priced at $87.00, is on sale at 12.5% off. Calculate the sale price.

e. A pen is priced at $120. Tax is added at 13%. Calculate the amount of tax.

f. A jacket is priced at $67.25. Tax is added at 12.5%. Calculate the amount of tax.

g. A book is priced at $52.80. Tax is added at 13%. Calculate the grand total.

h. A cell phone is priced at $520.00. Tax is added at 8%. Calculate the grand total.

i. A tie is priced at $25.99. It is on sale at 5% off. Tax is added at 15% of the sale price. Calculate the amount of tax.

j. A piano is priced at $5200. It is on sale at 12.5% off. Tax is added at 13% of the sale price. Calculate the amount of tax.

k. A notebook, priced at $14.80, is on sale at 10% off. Tax is added at 7% of the sale price. Calculate the grand total.

l. A sofa bed, priced at $550, is on sale at 15% off. Tax is added at 13% of the sale price. Calculate the grand total.

m. The total on the bill at the restaurant is $37.50. How much tip should you leave if you wish to leave a 15% tip?

n. The cab ride comes to $22.20. You would like to add a 15% tip. How much should you pay in total?

8. Solve the following problems using the algebraic technique.

a. 82.3% of those who graduate find employment within a year of graduation. How many graduates this year are expected to find employment by next year if there were 320 graduates this year?

b. 62.8% of the eligible voters voted. How many eligible voters voted if there are 12 430 eligible voters?

c. 42.8% of those who were polled picked Party A as their favourite party. How many of those who were polled did not pick Party A as their favourite party? A total of 1200 people were polled.

d. The population of the moose in the park decreased by 8.5% in the past year. How many moose are there in the park this year if there were 1820 moose in the park last year?

e. Production increased by 11.2% this year. How many units were produced this year if 13 000 units were produced last year?

f. Enrolment in the program increased by 7.9% this year. How many people enrolled in the program this year if 1300 people enrolled in the program last year?

g. 24.5% of those who participated in a clinical trial experienced side effects. Of these, 3.8% were hospitalized. How many participants who experience side effects are expected to be hospitalized in a study of 400 participants?

h. 22.3% of the club members left the club this year. Of those who left the club, 63% cited deterioration of the services as the reason for leaving. How many club members who left the club cited the deterioration of services as the reason for leaving if there were 1250 members in the club last year?

i. After the spillage of the chemicals in the river, 42.8% of the fish died. Of those that did not die, 61.9% reproduced. How many fish that did not die reproduced if there were 26 300 fish before the spillage?

j. 72.5% of the people voted in favour of amending the constitution. Of those who did not vote in favour of amending the constitution, 89.3% identified themselves as supporters of the ABC political party. How many of those who did not vote in favour of amending the constitution identified themselves as supporters of the ABC political party if 732 300 people voted?

k. 82.8% of the students in the program are part-time. Of those who are part-time, 78.5% take at least three courses. How many students in a class of 120 are expected to be part-time students, taking fewer than three courses?

l. 75.4% of those who were surveyed rated the subway service as being unsatisfactory. Of those who rated the subway service as unsatisfactory, 25.2% cited unreliability as the reason for rating the service as unsatisfactory. How many people rated the subway service as unsatisfactory for reasons other than unreliability if 1020 people were surveyed.

m. There are 370 eligible voters. 31.2% of them did not vote. Of those who voted, 61.2% rejected the motion. How many eligible voters voted and did not reject the motion?

n. 92.8% of adults of working age are employed. Of those who are unemployed, 42% are expected to find employment within six months. How many people who are unemployed are expected not to find employment in the next six months if there are 370 000 adults of working age?

9. Solve each of the following problems using the algebraic technique.

a. Find the conversion factor for converting the unit of mass from g to lb given that

$$1 \text{ kg} = 1000 \text{ g}$$
$$1 \text{ lb} = 0.4536 \text{ kg}$$

b. Find the conversion factor for converting the unit of time from s to h given that

$$1 \text{ h} = 60 \text{ min}$$
$$1 \text{ min} = 60 \text{ s}$$

c. Find the conversion factor for converting the unit of pressure from kPa to atm given that

$$1 \text{ Pa} = 0.001 \text{ kPa}$$
$$1 \text{ mmHg} = 133.322 \text{ Pa}$$
$$1 \text{ atm} = 760.0 \text{ mmHg}$$

d. In an outbreak of a virus, 19% of the population got sick. Of those who got sick, 2% were hospitalized. What percentage of the population got sick and were hospitalized?

e. 17.3% of the students who take a program do not graduate from the program. Of those who do graduate from the program, 21.3% graduate with a GPA below 2.00. What percentage of the students who take the program graduate with a GPA of at least 2.00?

f. $\frac{1}{8}$ of the budget is set aside for the customer service department. Of the rest, $\frac{1}{4}$ is given to the marketing department. What fraction of the budget is given to the marketing department?

10. Show that the expression $\frac{q_3 q_2 q_1}{q_4 q_5 q_6}$ is directly proportional to the subexpression $\frac{q_1}{q_4 q_6}$.

11. List the subexpressions of each of the expressions below that are directly proportional to the expression itself.

a. $q_3 q_2 q_1$

b. $\frac{q_2 q_1}{q_3}$

c. $\frac{q_1}{q_2 q_3}$

d. $\frac{1}{q_1 q_2 q_3}$

12. What does it mean when we say that direct proportion is symmetric?

Chapter 8
Inverse Proportion

If q_1 and q_2 represent the values of two quantities, then we say that q_2 is inversely proportional to q_1 if scaling q_1 by a certain factor[1] forces the scaling of q_2 by the inverse of the same factor.[2,3]

As noted in the chapter on inverse superposition, in the designation *inverse proportion* the adjective *inverse* is used to imply that the change in the size of q_2 is in the opposite direction to the change in the size of q_1, i.e., one increases in size while the other decreases in size, or one decreases in size while the other increases in size. The word *proportion* implies that the change takes place through multiplications and divisions. Note that a relationship may be inverse (in which case the size of the value of the one of the quantities increases while the size of the value of the other quantity decreases, or the value of one of the quantities decreases while the value of the other quantity increases) but not inverse proportion (the increase and decrease in the sizes of the two quantities are not related through a factor and its inverse). In a relationship of type *inverse proportion*, not only does the size of q_1 increase while the size of q_2 decreases, or the size of q_1 decreases while the size of q_2 increases, but the increase and decrease are related through a factor and its inverse.

---

[1]The phrase *scaling a value by a factor* means multiplying the value by that factor. As an example, scaling the value of mass by a factor of 2 means doubling the value of the mass and scaling the value of speed by a factor of $\frac{1}{2}$ means halving the value of the speed and so on.

[2]The phrase *scaling a value by the inverse of a factor* means multiplying the value by the inverse of that factor. As an example, scaling the value of length by the inverse of factor 2 means multiplying the value by $\frac{1}{2}$ (or, equivalently, dividing the value by 2) and scaling a value by the inverse of the factor $\frac{1}{3}$ means multiplying the value by 3 and so on.

[3]The definition of inverse proportion given above implies that multiplication/division of the value of one of the quantities by a number forces the division/multiplication of the value of the other quantity by the same number. As an example, multiplication of the value of one of the quantities by 2 (i.e., doubling its value), forces the division of the value of the other quantity by 2 (i.e., halves the value of the other quantity) and dividing the value of one of the quantities by 2 (i.e., halving its value) forces the multiplication of the value of the other quantity by 2 (i.e., doubles its value).

As an example of a relationship of type inverse proportion, consider the following everyday life problem.

Problem

5 painters paint a house in 4 days. At this rate, how long does it take 10 painters to paint the house?

An Algebraic Solution

t time it takes to paint the house (days)

$$t = \frac{5 \times 4}{10}$$

$$t = 2$$

It would take 2 days to paint the house.

Analysis of the expression on the right side of the equation that models the problem yields the single term $\frac{5 \times 4}{10}$ which breaks into the single factor $\frac{5 \times 4}{10}$. This factor involves a division with dividend 5×4 and divisor 10. The quotient, dividend and divisor of this division along with the quantities whose values they represent are listed below.

$\frac{5 \times 4}{10}$ the time it takes 10 painters to paint the house (days)

5×4 the time it takes 1 painter to paint the house (1·days)

10 number of painters (1)

This is illustrated below:

Dividend: number of days it takes
1 painter to paint the house

$$t \quad = \quad \frac{\boxed{5 \times 4}}{\boxed{10}} \quad \longleftarrow \quad \text{Quotient: number of days it takes}$$

Quotient: number of days it takes
10 painters to paint the house

Divisor: number of
painters

The equation that models the problem states that the *time it takes 10 painters to paint the house* is equal to the quotient of the *time it takes 1 painter to paint the house* and the *number of painters*.

In this problem the *time it takes to paint the house* is inversely proportional to the *number of painters*. Note that scaling the number of painters by a certain factor forces the scaling of the time it takes to paint the house by the inverse of the same factor. As an example, in the problem above 5 painters paint the house in 4 days. If there are twice as many painters, i.e., 10, then we would expect that the job would proceed at twice the speed so that the time it takes to paint the house would get halved to 2 days.

The quantity represented by a term is always inversely proportional to the inverses of quantities that are represented by the factors in it. If the expression is made up of a single term, then the quantity represented by the whole expression is the same as the quantity represented by the single term that it is made of and we can say that the quantity represented by the expression as a whole is inversely proportional to the quantities represented by the inverses of the factors within its single term. This was the case in the problem above. When the expression is made up of multiple terms, the quantity represented by each term is inversely proportional to the quantities represented by the inverses of the factors within it but the quantity represented by the whole expression is *not* inversely proportional to the inverses of any of the factors within any of its terms. The following problem illustrates the point.

Problem

You and your friend start driving toward two different destinations. Your destination is 400 km away and your friend's destination is 300 km away. You drive on a country road at a speed of 90 km/h. Your friend drives on a highway at a speed of 120 km/h. What is the difference between your arrival times?

An Algebraic Model

t the difference between our arrival times (h)

$$t = \frac{400}{90} - \frac{300}{120}$$

The equation above states that *difference between our arrival times* (represented by the term t) is equal to the difference between *my arrival time* (represented by the term $\frac{400}{90}$) and *my friend's arrival time* (represented by the term $\frac{300}{120}$).[4]

Analysis of the expression on the right side of the equation that models the problem into terms, factors, quotients, dividends and divisors yields the

---

[4]To work out travel time, one can divide the distance by travelling speed. As an example, it will take 2 h to cover a distance of 200 km at a speed of 100 km/h.

following for the second term on the right side of the equation.[5]

$$a \quad = \quad \frac{400}{90} \quad + \quad \frac{300}{120}$$

Dividend: distance my
friend covers
$\downarrow$

Quotient: my friend's
travel time

$\uparrow$
Divisor: my friend's
travelling speed

In this problem *my friend's travel time* is inversely proportional to *my friend's travelling speed*. Note that scaling my friend's speed by a factor of, say, $\frac{1}{2}$ (i.e., driving at half the speed), would double my friend's travelling time.

As we just noted, the quantity represented by the second term in the expression on the right side of the equation above, i.e., *my friend's travel time*, is inversely proportional to the quantities that the factors in its divisor represent, i.e., *my friend's travelling speed*. However, the quantity represented by the whole expression, i.e., *the difference between our arrival times*, is not inversely proportional to the quantities that the factors within the divisors of the quotients represent, i.e., *my travelling speed* and *my friend's travelling speed*. As an example, while doubling my friend's speed halves my friend's travel time (i.e., halves the value of $\frac{300}{120}$), it does not halve the difference between our arrival times (does *not* halve the expression $\frac{400}{90} - \frac{300}{120}$).

Let us now look at some of the applications of inverse proportion.[6]

Everyday Life Problems

Consider the following problem.

Problem

3 pumps can empty a swimming pool is 20 h. At this rate, how many pumps can empty the swimming pool in 4 h?

---

[5]We invite the reader to name the quantities that are represented by the first term, the factor within that term, and the quotient, dividend and divisor within that factor.

[6]In what follows we will use the standard model to setup our models. For an alternative approach using conservation formulations please see Appendix E.

We will present an algebraic solution for this problem. Following this, we will explain how the model for this problem is put together and the reason this setup works.

First we note that this is an example of an inverse proportion problem since the time it takes to empty the pool is inversely proportional to the number of pumps used. As an example, if we use twice as many pumps, it will take half as much time to empty the pool. Since the problem is of type *inverse proportion*, we can follow the guidelines below to model the problem.

To setup the expression on the right side of the equation that models the problem, it may help to make a note of the correspondence between the two values 3 pumps and 20 h as shown below.

3 pumps   20 h

This correspondence is given by the sentence 3 *pumps can empty a swimming pool in* 20 *h* in the body of the word problem.

Since this is an inverse proportion problem, we can setup the model as a division, i.e.,

$$\frac{3 \text{ pumps} \qquad 20 \text{ h}}{}$$

The values in the listed relationship will appear as factors in the dividend of this division. We write

$$\frac{3 \text{ pumps} \qquad 20 \text{ h}}{3 \times 20}$$

The divisor of this division is the value we wish to process: 4 h.

$$\frac{3 \text{ pumps} \qquad 20 \text{ h}}{\frac{3 \times 20}{4}}$$

The full solution without the intervening explanations is given below.

An Algebraic Solution

n number of pumps needed (1)

$$n = \frac{3 \times 20}{4}$$

$$n = 15$$

We need 15 pumps.

To understand how the setup above solves the problem, we draw the reader's attention to two facts.

First, the model for an inverse proportion problem relates the value of the quantity on the left (n, in the model above) and the value of the quantity in the divisor of the division on the right (4, in the model above). Variations of this problem may be generated by changing the value of 4 and noting its effect on n while keeping the type of pumps and the size of the swimming pool constant which, in effect, keeps the relationship between 3 pumps and 20 h fixed.

Second, the quantity in the dividend of the division on the right side (3×20, in the example above), plays a special role in the solution of problems of type *inverse proportion*. So long as the relationship between the factors remains the same (as in the relationship between 3 pumps and 20 h staying the same), their product represents the value of a quantity that remains the same and acts as a pivot that relates the products of the many possible combinations of *the number of pumps* and *the time it takes to empty the pool*.

In the example above, the dividend, i.e., 3×20, evaluates to 60. It is tempting to say that this is the total time that is needed to empty the pool. However, as we will see below, there is more to the semantics of this quantity than is apparent.

The numerical value of the dividend in the example above, i.e., 60, is generated through the product of the number of pumps (with unit 1) and the time it takes to empty the pool (with unit h). Here we prefer to think of it as stating that it takes 1 pump 60 h to empty the pool. We are now beginning to see how the model works: The numerator works out the time that it take 1 pump to empty the pool. The division by 4 indicates that the time that it takes 1 pump to empty the pool, i.e., 60, is divided into 4 h chunks. The number of chunks tells us how many pumps are needed.

To make sense of an equation of type inverse proportion, make sense of the dividend of the division on the right and then note the manner in which the division of this quantity by the divisor generates the value of the quantity on the left.

Problems from the Sciences

Many problems in the sciences involve quantities that are inversely proportional to each other. As an example of such relationships, consider the following problem from kinematics.

Problem

Travelling at a speed of 120 km/h, you can drive from Toronto to Montreal in 4.5 h. How long will it take to drive from Toronto to Montreal if you drive at 90 km/h?

The two quantities involved in the problem above are travelling speed and travel time. These quantities are related through inverse proportion: As an example, if you double your speed, you will cover the same distance in half the time, and, if you reduce your speed to one third of its value, it will take you 3 times as long to cover the same distance, and so on. Since this is an inverse proportion problem, we can use the guidelines given earlier to generate a model for it.

To begin, we note the related values 120 km/h and 4.5 h.

120 km/h   4.5 h

Since this is an inverse proportion problem, it can be modelled as a division. We write

120 km/h   4.5 h

$$t = \frac{}{}$$

The related values 120 km/h and 4.5 h appear as factors in the dividend of this division. We write

120 km/h   4.5 h

$$t = \frac{120 \times 4.5}{}$$

The value we wish to process, i.e., 90 km/h, appears as the divisor of this division. We write

120 km/h   4.5 h

$$t = \frac{120 \times 4.5}{90}$$

To see how the equation above models the problem, we draw the readers attention to the dividend and divisor of the division on the right side. As usual, in an inverse proportion problem the value of the quantity represented by the dividend of the division represents a value that is constant and has significance. In the example above, the dividend, 120×4.5 simplifies to 540 km. This represents the distance from Toronto to Montreal. Note that this distance is a constant. We now understand what this model is trying to achieve. To find travel time, the expression divides the distance from Toronto to Montreal (the dividend) by the desired travelling speed (the divisor).

The full solution to the problem above is given below.

An Algebraic Solution

t time it takes to drive to Montreal (h)

$$t = \frac{120 \times 4.5}{90}$$

$t = 6$ h

it will take 6 h to drive to Montreal.

Here is a problem involving the behaviour of ideal gases from thermodynamics.

Problem

A container of volume 6.5 m^3 contains helium under a pressure of 15 200 kPa. Calculate the pressure of the helium if its volume is reduced to 6.2 m^3. The temperature of the helium remains constant during the process.

Experimental data show that, in such a problem, the pressure of helium is inversely proportional to its volume.[7] We can, therefore, follow the guidelines given earlier to model this problem. We begin by listing the related values 15 200 kPa and 6.5 m^3.

15 200 kPa   6.5 m^3

Since the problem is of type *inverse proportion*, we can model the problem as a division. We write

15 200 kPa   6.5 m^3

$$p = \frac{\qquad}{}$$

The related values, 15 200 kPa and 6.5 m^3, appear as factors in the dividend of this division.

15 200 kPa   6.5 m^3

$$p = \frac{15\,200 \times 6.5}{}$$

The value that we wish to process, i.e., 6.2 m^3, appears as the divisor of this division.

15 200 kPa   6.5 m^3

$$p = \frac{15\,200 \times 6.5}{6.2}$$

---

[7]See the chapter on formulas for a more in-depth discussion of such systems.

As before, the value of the quantity represented by the dividend of this division is a constant that has some significance in the context of the word problem. For the present problem, it can be shown that the product $15\,200 \times 6.5$ represents the internal energy (heat content) of the system which, since the amount of helium and its temperature do not change in the process, remains constant. Here is the full solution.

An Algebraic Solution

p pressure of helium in tank (kPa)

$$p = \frac{15\,200 \times 6.5}{6.2}$$

$$p = 15\,935 \text{ kPa}$$

The helium has a pressure of $15\,935$ kPa.

Subexpressions

As stated earlier, in a given expression, the value of the quantity represented by a term is inversely proportional to the values of the quantities that are represented by the inverses of the factors within it. As an example, the expression

$$\frac{1}{q_1 q_2 q_3 q_4}$$

is inversely proportional to q_1, q_2, q_3 and q_4. Further to this, we can show that the value of the quantity represented by a term is inversely proportional to the values of the quantities that are represented by any combination of the inverses of its factors. As an example, in addition to being inversely proportional to q_1, q_2, q_3 and q_4, the expression $\frac{1}{q_1 q_2 q_3 q_4}$ is inversely proportional to $q_1 q_2$, $q_2 q_3$, $q_1 q_2 q_3$, $q_1 q_2 q_4$, etc., and of course the expression $q_1 q_2 q_3 q_4$. As an example of how this is the case, consider the subexpression $q_2 q_3$ in the divisor of this expression. We can show that $\frac{1}{q_1 q_2 q_3 q_4}$ is inversely proportional to $q_2 q_3$ by placing the subexpression within brackets which effectively turns the combo into a single factor in the divisor. This is shown below.

$$q = \frac{1}{q_1 \left(q_2 q_3\right) q_4}$$

The divisor of the expression on the right side now breaks into three factors: q_1, $q_2 q_3$, and q_4. It should now be apparent that $\frac{1}{q_4 q_3 q_2 q_1}$ is inversely proportional to $(q_2 q_3)$ since the subexpression is being divided by. Indeed, if $(q_2 q_3)$ increases in size by a factor of, say, 2, then we will be dividing by a value that is 2 times larger in size, resulting in a value for $\frac{1}{q_1 q_2 q_3 q_4}$ that is 2 times smaller in size and so on.

At times we may choose to name the quantity whose value is represented by an expression. As an example, we may choose to refer to the expression $\frac{1}{q_1 q_2 q_3 q_4}$ as q. This leads to the following equation:

$$q = \frac{1}{q_1 q_2 q_3 q_4}$$

Since q is the same as $\frac{1}{q_1 q_2 q_3 q_4}$, then any conclusions that we make about $\frac{1}{q_1 q_2 q_3 q_4}$ will hold true for q as well. As an example, in the equation above, we can say that the expression $\frac{1}{q_1 q_2 q_3 q_4}$ is inversely proportional to the subexpressions q_1, q_2, $q_3 q_4$, and so on. Since q is another way to refer to $\frac{1}{q_1 q_2 q_3 q_4}$, then we can say that q is inversely proportional to q_1, q_2, $q_3 q_4$, and so on.

When a term involves a division, we may choose to further analyze the factor into factors in its dividend and factors in its divisor. As an example, when we analyze the expression

$$\frac{q_3 q_2 q_1}{q_4 q_5 q_6}$$

we see the expression as a single term that contains a single factor. This factor may, however, be further analyzed into factors in its dividend, i.e., q_1, q_2, and q_3, and factors in its divisor, i.e., q_4, q_5, and q_6. It can now be shown that $\frac{q_3 q_2 q_1}{q_4 q_5 q_6}$ is inversely proportional to any of its subexpressions that consists of one factor (e.g., $\frac{1}{q_3}$, q_6, etc.) two factors (e.g., $\frac{1}{q_1 q_3}$, $\frac{q_6}{q_2}$, etc.), three factors (e.g., $\frac{1}{q_1 q_2 q_3}$, $\frac{q_6 q_4}{q_1}$, etc.), and so on, including the inverse of the full expression itself, i.e., $\frac{q_6 q_5 q_4}{q_1 q_2 q_3}$. See the exercises for a proof.

As before, we may choose to name the quantity whose value is represented by the full expression. Referring to the expression $\frac{q_3 q_2 q_1}{q_4 q_5 q_6}$ as q allows us to write

$$q = \frac{q_3 q_2 q_1}{q_4 q_5 q_6}$$

and to conclude that any conclusions that hold true about $\frac{q_3 q_2 q_1}{q_4 q_5 q_6}$ will hold true about q as well.

Finally, note that inverse proportion is symmetric. If the value of one quantity is inversely proportional to the value of another quantity, then the value of the second quantity is inversely proportional to the value of the first quantity. We will justify this claim in the pages ahead.

Exercise Set 8

1. Solve each of the following problems using the algebraic technique.

    a. 12 assembly lines can assemble the speakers in 18 days. How long will it take to assemble the speakers if we set up 15 assembly lines?

    b. It takes 8.5 h for 2 cleaners to clean an office. How long will it take for 6 cleaners to clean the office?

    c. Driving at a speed of 80 km/h, you can drive to Vancouver in 4.5 h. How fast would you have to drive to get to Vancouver in 4 h?

    d. Driving at a speed of 110 km/h, you can drive to Calgary in 1.8 h. How long will it take to get to Calgary if you drove at a speed of 90 km/h?

    e. A sample of air is inside a cylinder topped with a piston. The air is under a pressure of 340 000 Pa and has a volume of 0.75 m^3. The piston is pushed down until the volume of the air reduces to 0.62 m^3. What is the pressure of the air now? No air is allowed to enter or leave the cylinder and the temperature remains constant.[8]

    f. A sample of helium inside a tank has a volume of 3.95 m^3 and is under a pressure of 428 000 Pa. Calculate the volume of the tank if its pressure drops to 380 000 Pa. The tank is sealed and its temperature remains constant during the process.[9]

    g. The current through a wire with a voltage difference across it is inversely proportional to the resistance of the wire. A voltage difference across a wire with resistance 3.5 Ω sets up a current of 2.5 A through the wire. What current will flow through a wire with a resistance of 4.1 Ω if the same voltage difference is set across it?

    h. The current through a wire with a voltage difference across it is inversely proportional to the resistance of the wire. A voltage difference across a wire with resistance 3.5 Ω sets up a current of 2.5 A through the wire. What is the resistance of a wire if the same voltage difference sets up a current of 7.8 A through the wire?

    i. According to Newton's Laws, the acceleration of an object is inversely proportional to the mass of the object. A force, acting on an object of mass 20.9 kg, forces the object to accelerate at a rate of 2.45 m/s^2. What acceleration would such a force generate in an object with a mass of 32.8 kg?

---

[8]These conditions enforce a relationship of type inverse proportion between the pressure of the air in the cylinder and its volume.

[9]These conditions enforce a relationship of type inverse proportion between the pressure of the helium in the tank and its volume.

j. According to Newton's Laws, the acceleration of an object is inversely proportional to the mass of the object. A force, acting on an object of mass 12.8 kg, forces the object to accelerate at a rate of 1.53 m/s^2. The same force acting on another object generates an acceleration of 1.73 m/s^2. What is the mass of this other object?

2. Show that the expression $\frac{q_3 q_2 q_1}{q_4 q_5 q_6}$ is inversely proportional to the subexpression $\frac{q_6 q_4}{q_1}$.

3. List the subexpressions of each of the expressions below that are inversely proportional to the expression itself.

a. $\frac{1}{q_3 q_2 q_1}$ b. $\frac{q_2 q_1}{q_3}$ c. $\frac{q_1}{q_2 q_3}$ d. $q_1 q_2 q_3$

4. What does it mean when we say that inverse proportion is symmetric?

Part IV
Linearity

In this section we will discuss problems whose models involve a mix of superposition and direct proportions with the aim of calculating the values of one or more unknown factors belonging to different terms. Such problems are classified as linear problems. The label arises from the geometric representation of an equation of this type involving two unknown factors in different terms with geometric terminology used in tribute to the uniqueness and importance of this particular branch of math.

Chapter 9
Linear Equations in One Unknown

9.1 Introduction

Consider the following problem.

Problem

> You have two jobs. Your hourly rate of pay at the first job is
> $12.50/h and your hourly rate of pay at the second job is $15.10/h.
> Last week you worked 10 h at your first job and 8 h at your second
> job. How much money did you make last week?

One could follow the steps below to solve this problem in a step-by-step
manner.

1. Calculate the amount of money made from the first job by multiplying
   $12.50/h by 10 h to get $125.00.
2. Calculate the amount of money made from the second job by multiplying $15.10/h by 8 h to get $120.80.
3. Add the results of Steps 1 and 2 above to arrive at the total amount of
   money that you made last week, i.e., $245.80.

Alternatively, one can use the algebraic technique to model the problem
above as follows.

$$a \; = \; 12.50 \times 10 \; + \; 15.10 \times 8$$

One would then evaluate the terms (which evaluate to the monies made from
the first and the second jobs) and then add these to arrive at the total amount
of money made last week.

The algebraic technique is, of course, neater, more compact, more efficient, easier to troubleshoot and all that. However, there are deeper advantages to the use of the algebraic technique. One of these deeper advantages is that the use of the algebraic technique allows us to use the same logic to kickstart the solution to the same problem with a different unknown. To see how, consider the problem below which is identical to the one above except that it has a different unknown.

Problem

> You have two jobs. Last week you worked 10 h at your first job and 8 h at your second job. You made $245.80 in total. Your hourly rate of pay at your first job is $12.50/h. What is your hourly rate of pay at your second job?

The steps below solve this problem.

1. Calculate the amount of money made from the first job by multiplying $12.50/h and 10 h to arrive at $125.00.
2. Subtract this from the total money that you made last week to find how much you made at your second job last week. This yields $120.80.
3. Divide this by hours worked at your second job last week to find your hourly rate of pay at your second job. This evaluates to $15.10/h.

The steps employed in solving this problem are very different from those used to solve previous problem. This is a setback as it seems that one would have to follow a different logic even for problems that are so similar. And it is here where we see a key advantage of using algebra to solve word problems as it allows us to use the same logic to model both problems. For both problems above, we begin by noting that *total money made equals to the sum of monies made at the two jobs*, regardless of what the unknown quantity is. This yields the following model for the second problem above.

$$245.80 = 12.50 \times 10 + r \times 8$$

One would then manipulate this equation until one finds the value of r.

The ability to kickstart the solution to problems that are identical except for the unknown by using the same logic to arrive at a model deals with a major concern when solving a word problem: Where do I start? *Start by relating the values of the various quantities involved in the word problem using the simplest possible logic.*

One might, of course, note that while it *is* true that we can employ the same logic to model problems that are identical except for the unknown, we still have to *solve* the resulting model, i.e., calculate the value of the unknown quantity. So the ease in kick-starting the solution afforded to us by the algebraic technique is offset by the need to solve the resulting model. In addition, one might note that while the use of algebra seems to ease the task of solving the problem, it tells us nothing about the logic behind the solution *the way a step-by-step solution does.* We will address these concerns in the sections ahead. For now, we simply address the first concern by noting that the steps involved in solving such models are few and quite straightforward and we respond to the second concern by noting that, provided semantic tools are used in solving the model, the computations in the solution of algebraic models can be mapped onto a set of steps for a step-by-step solution with relative ease. As such, not only are algebraic solutions able to expose the steps in step-by-step solutions, so long as one can model the problem and solve the resulting model, such solutions can be used as a self-teaching tool in explaining how to solve an unfamiliar problem using a step-by-step approach.

Another major advantage of the use of the algebraic technique is in its ability to answer the question as to whether enough information is given to solve a given problem for an unknown. If the model involves a single equation that contains a single unknown, then this is so otherwise it is not.

9.2   Linear Models

A problem is linear if its model analyzes into a superposition involving one or more terms each of which analyzes into a direct proportion involving one or more factors with the unknowns appearing only once as factors in different terms.[1]

The models for the problems presented in the previous section are both linear in their respective unknowns. In the model for the first problem, reproduced below, the unknown, a, representing the total income, appears as a factor in the term on the left side of the equation.

$$a \ = \ 12.50 \times 10 \ + \ 15.10 \times 8$$

We say that this problem is linear in the unknown a. In the model for the second problem, reproduced below, the unknown, r, representing the hourly rate of pay at the second job, appears as a factor in the second term on the right side of the equation.

$$245.80 \ = \ 12.50 \times 10 \ + \ r \times 8$$

---

[1]If the unknown belongs to a term within a factor, a situation that may arise when brackets are present, then the same criteria may be applied to the expression within that factor.

We say that the equation is linear in the unknown r.

In this section we will focus on the skill to model a variety of linear problems. In the next section we will show how linear models are solved.

The first step in modelling any problem is to determine whether the quantities involved in the problem can be related through superposition, i.e., whether certain quantities add and/or subtract to generate other quantities. The problems presented in the previous section, as an example, fall into the category of superposition: One argues that total income equals to the sum of the incomes at the two jobs.

Once we have considered superposition, then we consider each term involved in the equation and see whether each can be solved through proportion. In the problems presented in the previous section, as an example, one can argue that income from the first job equals to the product of the hourly rate of pay at the first job and the number of hours worked at the first job. This is a proportion problem. A similar argument applies to the calculation of the income from the second job.

Problem

The mass of a molecule of $X_6H_{12}O_6$, is 180.156 amu. Identify the unknown atom, X, by finding its atomic mass. The atomic masses of H and O are 1.008 amu/atom and 16.00 amu/atom respectively.

The simplest way to model this problem is to argue that the mass of a molecule of $X_6H_{12}O_6$ is equal to the sum of the masses of the X, H and O atoms in that molecule. This is the superposition stage.

Next we argue that the mass of X atoms is equal to the product of the atomic mass of X and the number of X atoms. Similarly, the mass of H atoms is equal to the product of the atomic mass of H and the number of H atoms. And the mass of the O atoms is equal to the product of the atomic mass of O and the number of O atoms. We have the following model.

$$180.156 = M \times 6 + 1.008 \times 12 + 16.00 \times 6$$

where M represents the atomic mass of X in amu/atom.

Here is a problem involving brackets.

Problem

The mass of 2.8 mol of CX_2 is 123.228 g. Identify the unknown atom, X, by finding its molar mass. The molar mass of C is 12.01

g/mol.

The simplest argument that leads to a model for this problem is that the mass of 2.8 mol of CX_2 is equal to 2.8 times the mass of 1 mol of CX_2. Note that this problem is not superposition as the mass of 2.8 mol of CX_2 can be obtained through *multiplication* not *addition/subtraction*. The subproblem involving the determination of the mass of 1 mol of CX_2 *is*, on the other hand, a superposition of the mass of a C atom and the mass of 2 X atoms. The model follows.

$$123.228 = 2.8\,(12.01 + M \times 2)$$

where M represents the molar mass of X in g/mol.

Average problems can always be modelled as sum of data divided by the number of data items.

Problem

You wrote three tests of equal weight. You scored 70% on Test 1 and 80% of Test 2. What was your score on Test 3 if your average after Test 3 was 72%?

The simplest logic to employ to solve this problem is that the average mark is equal to the sum of the grades divided by the number of grades. This yields the following model.

$$72\% = \frac{70\% + 80\% + g}{3}$$

where g represents the grade on Test 3.

Here is a weighted average problem.

Problem

Test 1 is worth 10% of the final mark. Test 2 is worth 5% of the final mark. Test 3 is worth 20% of the final mark. You scored 80% on Test 1 and 60% on Test 2. What mark on Test 3 will give you an average of 75%?

The model for this problem is given below.

$$75\% = \frac{10 \times 80\% + 5 \times 60\% + 20 \times m}{10 + 5 + 20}$$

The following alternative that uses the decimal interpretation of percentage is generally preferred.

$$75\% = \frac{0.1 \times 80\% + 0.05 \times 60\% + 0.2 \times m}{0.1 + 0.05 + 0.2}$$

Here is a weighted average problem involving atomic mass.

Problem

You have a sample of oxygen from the lab. The sample contains 95.8% ^{16}O. The remaining 4.2% is an unknown isotope of O. Identify the unknown isotope of O in the sample if the average mass of an O atom in the sample is 16.084 amu.

Here is a model for this problem.

$$0.958 \times 16 + 0.042 \times m = 16.084$$

Here is a money problem involving discount and tax.

Problem

A hat, priced at $15.20, is on sale. Tax is added at 13%. Find the rate of discount if the grand total comes to $14.60.

We can model this problem as follows.

$$14.60 = (1 + 0.13)(1 - r)(15.20)$$

where r represents the rate of discount.

For problems that involve a fraction, decimal fraction or percent of a value, we reason by noting that a certain piece is equal to a fraction, decimal fraction or percent of that value. Here is an example.

Problem

16.5% of the frogs in the sample have red eyes. How many frogs are there in the sample if 1048 frogs in the sample do *not* have red eyes?

A model for this problem is given below.

$$1048 = (1 - 0.165) n$$

where n represents the number of non-red-eyed frogs in the sample.

Exercise Set 9.2

Model each of the following problems as a linear equation in one unknown.

1. On average we can extract 4 Cal/g of carbohydrate, 9 Cal/g of fat and 4 Cal/g of protein. A glass of milk contains 12.5 g of carbohydrate, 1.4 g of fat and an unknown mass of protein. What mass of protein is there in the glass of milk if its energy content is 79.4 Cal?

2. On average we can extract 4 Cal/g of carbohydrate and 9 Cal/g of fat. A cup of yogurt contains 16 grams of carbohydrate, 4.2 grams of fat and 15.7 g of protein. The energy content of the cup of yogurt is 164.6 Cal. What is the energy content per g of protein?

3. For breakfast Marco had 3 slices of buttered bread at 37 Cal/slice, 2 spoons of jam at 52 Cal/spoon and a cup of tea. He rested for 2 hours, burning on average 80 Cal/h. Then he exercised for 0.5 h, burning Calories at a rate of 320 Cal/h. What is the energy content of the cup of tea if Marco's net intake of energy since breakfast is −85 Cal?

4. You bought 6 pens at $2.50/pen and 8 notebooks. What was the cost of a notebook if you spent $54.92 in total?

5. You bought 6.5 metres of a fabric at $1.20/m and a few rolls of string at $2.00/roll. How many rolls of string did you buy if you spent $13.80 in total?

6. You order 20 boxes of printer paper. Each box contains a number of packages of paper at $2.99/package. How many packages are there in each box if in total you spent $1076.40.

7. You order 25 cartons of masks. Each carton contains 12 boxes of masks. What is the cost of a box of masks if the order costs $960?

8. You order 200 pens at $1.50/pen and 15 boxes of printer paper. Each box contains 8 packages of printer paper. What is the cost of a package of printer paper if the order costs $538.80?

9. You bought a calculator for $16.99 and 3 pens at $2.00/pen. How much did you start with if you have $35.01 left?

10. You start with a certain volume of medication. You use 5.65 ml/min of the medication for 12 minutes. What volume of the medication did you start with if you are left with 332.2 mL of the medication?

11. You start with 180 mL of a medication. You use 2.4 mL/min of the medication for some time. How long did you use the medication if you are left with 108 mL of the medication?

12. You start with 180 mL of a medication. You use the medication at a constant rate for 30 min. At what rate did you use the medication if you are left with 108 mL of the medication?

13. The mass of a molecule of C_2X_6 is 30.068 amu. Identify the unknown atom, X, by finding its atomic mass. The atomic mass of C is 12.01 amu/atom.

14. The mass of 1 mole of X_2H_6 is 30.068 g. Identify the unknown atom, X, by finding its molar mass. The molar mass of H is 1.008 g/mol.

15. The mass of one molecule of C_nH_6 is 30.068 amu. How many atoms of C are there in a molecule of C_nH_6? The atomic masses of C and H are 12.01 amu/atom and 1.008 amu/atom respectively.

16. The mass of 42 molecules of SX_2 is 2690.94 amu. Identify the unknown atom, X, by finding its atomic mass. An atom of S has a mass of 32.07 amu.

17. The mass of 42 moles of XO_2 is 2690.94 g. Identify the unknown atom, X, by finding its molar mass. The molar mass of O is 16.00 g/mol.

18. The mass of a molecule of $Ca(XH)_2$ is 74.096 amu. Identify the unknown atom, X, by finding its atomic mass. An atom of Ca has a mass of 40.08 amu. The atomic mass of H is 1.008 amu/atom.

19. The mass of 1 mol of $X_2(CO_3)_4$ is 253.922 g. Identify the unknown atom, X, by finding its molar mass. The molar masses of C and O are 12.01 g/mol and 16.00 g/mol respectively.

20. The mass of 4.8 moles of $Na_4(XO_4)_3$ is 1808.976 g. Identify the unknown atom, X, by finding its molar mass. The molar masses of Na and O are 22.99 g/mol and 16.00 g/mol respectively.

21. A sample of $(NH_4)_3PO_4$ has a mass of 15 059.386 amu. The atomic masses of N, H, P and O are 14.01 amu/atom, 1.008 amu/atom, 30.97 amu/atom and 16.00 amu/atom respectively. How many molecules of $(NH_4)_3PO_4$ are there in the sample?

22. The average temperature over the past 5 days was 15.8 °C. What was the temperature on the third day if the temperature on the first, second, fourth and fifth days were 12 °C, 15 °C, 14 °C, and 20 °C?

23. On average, Marie spends $2000/month. Over the past two months she spent $2100 and $1600. What is the maximum amount of money she can spend this month if she wishes to maintain her average monthly expense?

24. Fiona ordered 8 Type A computer systems at $1200/system, a Type B computer system for $820, and 3 Type C computer systems. What is

the cost of a Type C computer system if the average cost of a system in the order was $1043.33?

25. Last week you worked 15 hours at a job earning $14/hour, 10 hours at another job earning $16/hour and 2 hours at a third job. What is your hourly rate of pay at your third job? Your average hourly rate of pay last week was $15.04/h.

26. There are 120 workers in Section A, 250 workers in Section B, and 320 workers in Section C. Workers in Section A receive 3 hours of training per week. Workers in Section B receive 5 hours of training per week. How many hours of training per week are received by a worker in Section C if the average training time per week per worker across the the three sections is 4 h/week/worker?

27. You took three semester-long courses: MATH101, BIOL101 and CHEM101 with credit values of 3, 5 and 4 respectively. You received a final letter grade of B+ in MATH101 worth 3.33 grade points and a C− in BIOL101 worth 1.67 grade points. What grade point in CHEM101 will give you a GPA of 3.00?

28. You took PHOT1148, PSY1040, GHUM1014 and COMM1020 each of which was one semester long. The contact hours for the four courses were 3 h/week, 2 h/week, 4 h/week and 3 h/week respectively. You received an A in PHOT1148 worth 4.00 grade points, a B− in PSY1040 worth 2.67 grade points and a C+ in GHUM1014 worth 2.33 grade points. What grade in COMM1020 will give you a GPA of 3.40? Each contact hour counts as a credit value.

29. A hat is on sale at 20% off. Find the price of the hat if the amount of discount is $4.50.

30. A soccer ball, priced at $18.20 is on sale. Calculate the rate of discount if the sale price comes to $16.38.

31. You bought a history textbook and paid $3.38 in taxes. What was the price of the textbook if tax is added at 13%?

32. A pair of gloves is on sale at 15% off. Tax is added at 13%. Find the price if the grand total comes to $17.48?

33. A book, priced at $18.20, is on sale at 15% off. Tax is added to the sale price. Find the rate of tax if the grand total comes to $17.48?

34. A laptop is on sale at 15% off. Tax is added at 13% of the sale price. Find the price of the laptop if you paid $132.60 in taxes?

35. A pen, priced at $24.80, is on sale. Tax is added at 13%. Find the rate of discount if the amount of tax is $2.90.

36. The cost of your ride is $19.10. You gave the driver a $3.00 tip. What percentage of the cost of the ride is the tip?

37. 82% of the respondents chose soft drink A as their favourite soft drink. How many respondents were there if 1246 respondents chose soft drink A as their favourite soft drink?

38. 82% of the respondents chose soft drink A as their favourite soft drink. How many respondents were there if 274 respondents did not choose soft drink A as their favourite soft drink?

39. The population of a town increased by 5544 people over the past ten years. What is the percent rise in the population of the town if the population was 132 000 ten years ago?

40. The popultion of a town increased by 5544 people over the past ten years. This represents a growth of 4.2%. What was the population ten years ago?

41. 78% of the eligible voters voted. Of these 18% opposed the proposed changes. How many eligible voters are there if 76 eligible voters who voted opposed the proposed changes?

42. 78% of the eligible voters voted. Of these 18% opposed the proposed changes. How many eligible voters are there if 346 eligible voters who voted did not oppose the proposed changes?

43. 18% of the eligible voters who voted opposed the proposed changes. What percentage of the eligible voters voted if 76 eligible voters who voted opposed the proposed changes? There are 541 eligible voters in total.

44. 18% of the eligible voters who voted opposed the proposed changes. What percentage of the eligible voters voted if 346 eligible voters who voted did not oppose the proposed changes? There are 541 eligible voters.

45. 78% of the eligible voters voted. Of these, some opposed the proposed chanegs. What percentage of eligible voters who voted opposed the proposed changes if 76 eligible voters who voted opposed the proposed changes? There are 541 eligible voters.

46. 78% of the eligible voters voted. Of these, some opposed the proposed chanegs. What percentage of eligible voters who voted opposed the proposed changes if 346 eligible voters who voted did not oppose the proposed changes? There are 541 eligible voters.

47. 34% of the frogs in the sample have green skin. Of those that have green skin, 15% have red eyes. How many frogs are there in the sample if there are 62 red-eyed, green-skinned frogs in the sample?

48. In a recent study, 61% of the participants experienced side effects. Of those who experienced side effects, some had to be hospitalized. What percentage of those who experienced side effects needed to be hospitalized if 70 participants who expereinced side effects did not have to be

hospitalized. There were 120 participants. in the study.

49. Blue Lake City had a population of 180 000 at the begining of eighties. The population of Blue Lake City increased by 12% in the eighties and then decreases in the nineties. What was the rate of decrease in the population of Blue Lake City in the nineties if it had a population of 181 440 at the end of nineties?

50. Statistics show that 33% of those who commute to work daily take the public transit system. Of those who do not take the public transit system, 72% drive to work. How many people commute to work daily if 45 000 people who commute to work daily and do not take the public transit system do not drive to work?

9.3   Solution Techniques

As we noted in the previous sections, finding a model for a problem is not the end of the solution process in the algebraic approach. One must still *solve* the resulting model for the unknown. This means that the resuling mathematical model should be logically manipulated until one isolates the unknown, i.e., until one arrives at an equation that has the form of

$$q = \ldots$$

where q represents the unknown value of the quantity we wish to work out.

A number of algorithms and alternatives exist for solving different types of equations. In this textbook we will use the following algorithm to solve linear models where the unknown appears only once throughout the equation.[2,3]

---

[2]This algorithm is discussed in detail in the companion textbook *Style in Technical Math*. The algorithm can be extended with ease to deal with linear equations in which the unknown appears multiple times. A different extension generates a solution algorithm for nonlinear problems in which the unknown apprears only once. Nonlinear equations in which the unknown appears multiple times and cannot be paraphrased as an equation with the unknown appearing only once are not covered in this textbook.

[3]Since linear equations are a mix of superposition and direct proportion, it makes sense that by working with entities that are being added or subtracted (terms) and multiplied (factors) we should be able to solve such problems.

It is also interesting to note that linear equations involve unknowns that appear as a term, a factor within a term, a term within a factor within a term, etc. Therefore, it is by working with the terms and the factors that we should be able to solve them.

Algorithm for Solving Linear Equations with the Unknown Appearing Once Throughout the Equation

1. Solve for the term that contains the unknown.
2. Solve for the factor that contains the unknown.
3. Repeat Steps 1 and 2 until the unknown is isolated.

Different logical schemes can be used to take the steps in the algorithm above. In this section we will discuss two such logical schemes. The first uses the kind of logic that converts the main operation on one side of an equation to its inverse operation on the other side of the equation. This logic is aligned with the way we naturally reason and, for this reason, we refer to schemes that use this kind of logic as **natural semantic schemes**. The second uses the kind of logic that allows one to do the same thing to both sides of an equation. Following this logic, one views an equation as a balance and, if the balance holds, one can add equal amounts to both sides of the balance and still keep the balance, or multiply both sides by the same amount (e.g., double both sides, triple both sides, etc.) and still keep the balance, etc. We refer to schemes that use this kind of logic as **standard semantic schemes**.

Further to the above, each scheme can be implemented with or without intermediate simplifications giving rise to the following four flavours:

1. The natural semantic scheme without intermediate simplifications
2. The natural semantic scheme with intermediate simplifications
3. The standard semantic scheme without intermediate simplifications
4. The standard semantic scheme with intermediate simplifications

As we will see below, natural semantics schemes are more meaningful and more efficient than standard semantic schemes. In addition, as we will see below, those approaches that do not permit intermediate simplifications are best able to capture the logic that flows through the solution.

This implies that the best option in the list above is the first one: The natural semantic scheme without intermediate simplifications.[4] The second option, the natural semantic scheme with intermediate simplifications, is somewhat more efficient and easier to work with but the simplicity comes at a big price as intermediate simplifications can easily cloud the argument that flows through the solution and at times lead to logical steps that, while sound, follow a path that one would not naturally follow when dealing with everyday life problems.

While we do not recommend the use of the standard semantic scheme in working with equations, we do recommend familiarity with the approach as it

---

[4]It is also the only approach that can be easily adapted for use in rearranging formulas. See the chapter on formulas for more on this.

is used by some and familirity with the approach allows one to communicate with such individuals. The flavour used is the one that permits intermediate simplifications.[5]

To illustrate the use of the different approaches listed above, we present an example.

Problem

Last week you worked for 10 h at your first job and 8 h at your second job. What is your hourly rate of pay at your first job if you made $245.80 in total? Your hourly rate of pay at your second job is $15.10/h.

A step-by-step solution for this problem is given below.

A Step-by-Step Solution

1. Calculate the amount of money made at your second job.

$$15.10 \times 8 \ = \ 120.80$$

I made $120.80 at my second job.

2. Calculate the amount of money made at your first job.

$$245.80 \ - \ 120.80 \ = \ 125.00$$

I made $125.00 at my first job.

3. Calculate your hourly rate of pay at your first job.

$$125.00 \div 10 \ = \ 12.50$$

My hourly rate of pay at my first job is $12.50/h.

---

[5]The flavour that does not permit intermediate simplifications requires a level of mastery that is often not present in individuals who use standard semantics. For this reason, it is quite rare to see someone use this flavour. A hybrid scheme can be devised with standard semantics that allows for partial simplifications to lessen the complexity of the expressions that would otherwise inevitably result from the use of this approach with a full ban on intermediate simplifications. We will not present the details of this modified approach as it is still quite inefficient and, from the point of view of semnatics, inferior to its counterpart that uses natural semantics.

For the algebraic technique, we begin by modelling the problem following the simplest logic that relates the various quantities in the word problem. This logic states that total amount of money made is equal to the sum of the amounts of monies made from the first and second jobs, i.e.,

$$r \quad \text{rate of pay at first job (\$/h)}$$
$$245.80 = r \times 10 + 15.10 \times 8$$

We now show how this model can be solved using each of the four flavours above.

Natural Semantics without Intermediate Simplifications

Following natural semantics without intermediate simplifications, we follow the steps in the algorithm above but do not perform any calculations along the way until the unknown is isolated. We refer to this stage as the logical stage. Following the logical stage we begin the process of evaluating the expression on the right side of the equation. This is the simplification stage. The solution ends when the expression on the right side evaluates to a value.[6]

To see natural semantics without intermediate simplifications in action, we present the solution to the model above. We begin by switching sides in the model to bring the unknown to the left side of the equation. This yields

$$r \times 10 + 15.10 \times 8 = 245.80$$

This convention aids later in the interpretation of the logic that leads from one line of the solution to the next.

Following the algorithm above, we begin by solving the equation for the term that contains the unknown. This is the term $r \times 10$ and, by solving for it, we mean having the term as the only expression on the left side of the equation. To isolate this term we need to get rid of the addition of the term 15.10×8 on the left side. Following natural semantics, we convert the addition of the term 15.10×8 on the left side of the equation to subtraction of the same term 15.10×8 on this right side of the equation. This move is illustrated below.

$$r \times 10 + 15.10 \times 8 = 245.80$$
$$r \times 10 = 245.80 - 15.10 \times 8$$

Justification for this move can be given on semantic grounds: If *the sum of the monies made form my two jobs is equal to the total amount of money made*, i.e.,

$$r \times 10 + 15.10 \times 8 = 245.80$$

---

[6]Or else turns out to be indeterminate or undefined.

then *the income from the first job equals to the total income minus income from the second job*, i.e.,

$$r \times 10 = 245.80 - 15.10 \times 8$$

The logic above can be generalized as follows: If the sum of two things is equal to a total, then one of those things is equal to the total minus the other. This means that one can always turn an equation involving the addition of a term into a logically equivalent equation where the term is subtracted on the other side of the equation.[7,8]

We return to the algorithm above and continue by solving the equation for the factor that contains the unknown. This is the factor r and, by solving for it, we mean having the factor as the only expression on the left side of the equation. To isolate this factor we need to get rid of the multiplication by the factor 10 on the left side of the equation. Following natural semantics, we convert the multiplication by the factor 10 on the left side of the equation to division by the same factor 10 on the right side of the equation. This move is illustrated below.

$$r \times 10 = 245.80 - 15.10 \times 8$$
$$r = \frac{245.80 - 15.10 \times 8}{10}$$

Justification for this move can be given on semantic grounds: If *the product of the hourly rate of income at the first job and time worked at the first job equals to money made at the first job*[9], i.e.,

$$r \times 10 = 245.80 - 15.10 \times 8$$

then *the hourly rate of income at the first job equals to money made at the first job divided by time worked at the first job*, i.e.,

$$r = \frac{245.80 - 15.10 \times 8}{10}$$

---

[7]A different view sees the move from addition of a term on one side of the equation to the subtraction of that term on the other side of the equation as forward and backward steps. See the companion textbook *Style in Technical Math* for a detailed discussion of this view.

[8]The logic above can be used in reverse to convert the subtraction of a term on one side of an equation to addition of that term on the other side of the equation. As an example, we could argue that if *the income from the first job is equal to total income minus income from the second job*, i.e.,

$$r \times 10 = 245.80 - 15.10 \times 8$$

then *the sum of the incomes from the two jobs is equal to total income*, i.e.,

$$r \times 10 + 15.10 \times 8 = 245.80$$

[9]With the money made at the first job expressed as the difference between total money made and money made from the second job.

This argument can be generalized as follows: If the product of two things is equal to some product, then each of those things is equal to the product divided by the other. This means that one can always turn an equation involving the multiplication of a factor into a logically equivalent equation where the factor divides the other side.[10],[11]

The solution so far is summarized below.

r rate of pay at first job ($/h)

$$245.80 = r \times 10 + 15.10 \times 8$$
$$r \times 10 + 15.10 \times 8 = 245.80$$
$$r \times 10 = 245.80 - 15.10 \times 8$$
$$r = \frac{245.80 - 15.10 \times 8}{10}$$

The isolation of r signals the end of the logical stage. Note that the partial solution above does yet include any actual calculations. What we have done so far is to use the scheme that allows the conversion of the main operation on one side of an equation to its inverse on the other side of the equation to successively solve for the term that contains the unknown and the factor that contains the unknown. Actual calculations begin now.

$$r = \frac{245.80 - 15.10 \times 8}{10}$$
$$r = \frac{245.80 - 120.80}{10}$$
$$r = \frac{125.00}{10}$$
$$r = \$12.50/h$$

This states that my hourly rate of pay at my first job is $12.50/h.

---

[10]A different view sees the move from multiplication by a factor on one side of the equation to the division by that factor on the other side of the equation as forward and backward steps. See the companion textbook *Style in Technical Math* for a detailed discussion of this view.

[11]The logic above can be used in reverse to convert the division by a factor on one side of an equation to multiplication by that factor on the other side of the equation. As an example, we could argue that if *the hourly rate of income is equal to money made from the first job divided by time worked at the first job*, i.e.,

$$r = \frac{245.80 - 15.10 \times 8}{10}$$

then *the product of the hourly rate of income at the first job and time worked at the first job equals to money made at the first job*, i.e.,

$$r \times 10 = 245.80 - 15.10 \times 8$$

A key advantage of postponing the calculations is that the rearranged equation, i.e.,

$$r = \frac{245.80 - 15.10 \times 8}{10}$$

can serve as a guide to teach us how to solve the problem in a step-by-step manner. The steps in a step-by-step solution can be extracted from this rearranged equation by noting the sequence of calculations that need to be performed on the right side of the equation.[12] In the expression above, to work out the expression on the right side of the equation, one would begin by working out the subexpression 15.10×8. A bit of reflection shows that this finds the amount of money made from the second job as it multiplies my hourly rate of income at the second job by time worked at the second job and corresponds to Step 1 in the step-by-step solution given earlier. We arrive at

$$r = \frac{245.80 - 120.80}{10}$$

Next, we would perform the subtraction $245.80 - 120.80$. A bit of reflection shows that this finds the amount of money that I made at the first job as it involves subtracting the amount of money that I made at my second job from the total amount of money that I made and corresponds to Step 2 in the step-by-step solution given earlier. We arrive at

$$r = \frac{125.00}{10}$$

Finally, we would perform the division $\frac{125.00}{10}$. This computes my hourly rate of pay at the first job as it involves the division of the amount of money that I made at my first job by time worked at my first job and corresponds to Step 3 in the step-by-step solution given earlier. We arrive at

$$r = \$12.50/h$$

The ability to extract the steps in a step-by-step solution from an algebraic solution is a key skill whose mastery is highly recommended. So long as the reader is able to model the problem and rearrange the equation for the unknown using the semantic scheme without intermediate simplifications, the reader can extract the steps in a step-by-step approach by noting the sequence of calculations that are performed in evaluating the expression on the right side of the rearranged equation.[13] This is a powerful self-teaching tool and one that can be used with confidence.

---

[12] Returning to our earlier theme of analyzing and synthesizing equations, the present activity traces the steps in the synthesis of the expression on the right side of the equation.

[13] Note that by counting the number of operations in the rearranged equation we can tell how many steps we should expect in the corresponding step-by-step solution even before

Postponing the calculations in the solution also make it easier to keep account of the manner in which rounding rules in science problems are enforced. And as we will see later in the text, it also naturally generalizes to the study of formulas where intermediate simplifications are generally not possible.

Here is the full solution to the problem above without the intervening comments.

An Algebraic Solution using Natural Semantics without Intermediate Simplifications

r rate of pay at first job (\$/h)

$$245.80 = r \times 10 + 15.10 \times 8$$

$$r \times 10 + 15.10 \times 8 = 245.80$$

$$r \times 10 = 245.80 - 15.10 \times 8$$

$$r = \frac{245.80 - 15.10 \times 8}{10}$$

$$r = \frac{245.80 - 120.80}{10}$$

$$r = \frac{125.00}{10}$$

$$r = \$12.50/\text{h}$$

My rate of pay at the first job is \$12.50/h.

We now show how the problem above can be solved using natural semantics with intermediate simplifications.

Natural Semantics with Intermediate Simplifications

Following natural semantics with intermediate simplifications, we follow the steps in the algorithm above but allow for the simplification of the equations[14] that result from the execution of the steps in the algorithm. Simplification preceeds the first step in the algorithm and is performed after each step including the last step.

---

working the steps out. As an example, since the rearranged equation

$$r = \frac{245.80 - 15.10 \times 8}{10}$$

involves three operations (multiplication, subtraction and division), the corresponding step-by-step solution will involve three steps.

[14]By *simplifying an equation* here we mean executing any additions, subtractions, multiplications, etc., in that equation that can be performed.

Unlike natural semantics without intermediate simplifications which neatly separates the logical and computational steps, the use of this approach results in an interlacement of the logical and computational steps which is a disadvantage as one cannot tell at a glance whether a given line results from taking a logical step or a computational step. Furthermore, it becomes more difficult now to extract the steps in the corresponding step-by-step solution as the steps are scattered throughout the solution. The same can be said about tracking the rounding of the simplified values in the solution of applications of scientific theories. And, as the solution proceeds, intermediate values keep appearing, repeating, and vanishing, making it more and more difficult to understand what is happening at each stage. The most limiting aspect of this approach, however, is its inability to connect well with the schemes that are used to rearrange formulas. See the chapter on formulas for more on this.

The approach is, however, slightly more efficient than the one that does not permit intermediate simplifications and is easier to use as it requires the movement of a single numerical value at each step as opposed to nontrivial expressions.

To see how natural semantics with intermediate simplifications works, we present the solution to the problem aobve following this approach. As before, we begin by switching sides in the model to get

$$r \times 10 \ + \ 15.10 \times 8 \ = \ 245.80$$

We now simplify the equation by working out the term 15.10×8 to get

$$r \times 10 \ + \ 120.80 \ = \ 245.80$$

Following the algorithm given above, we solve for the term that contains the unknown.

$$r \times 10 \ = \ 245.80 \ - \ 120.80$$

We now simplify the equation by evaluating the expression on the right side. This yields

$$r \times 10 \ = \ 125.00$$

We continue with the algorithm above and solve for the factor that contains the unknown.

$$r \ = \ \frac{125.00}{10}$$

We now simplify the equation by working out the division on the right side.

$$r \ = \ \$12.50/h$$

My rate of pay at my first job is $12.50/h.

The full solution without the intervening comments is given below.

An Algebraic Solution using Natural Semantics with Intermediate Simplifications

r my hourly rate of pay at my first job (\$/h)

$$r \times 10 \ + \ 15.10 \times 8 \ = \ 245.80$$

$$r \times 10 \ + \ 120.80 \ = \ 245.80$$

$$r \times 10 \ = \ 245.80 \ - \ 120.80$$

$$r \times 10 \ = \ 125.00$$

$$r \ = \ \frac{125.00}{10}$$

$$r \ = \ \$12.50/\text{h}$$

My rate of pay at my first job is \$12.50/h.

We now move on to explain the manner in which the problem above can be solved using standard semantics with intermediate simplifications.[15]

Standard Semantics with Intermediate Simplifications

An alternative to the natural semantic scheme which is based on the logic that the main operation on one side of an equation can be converted to its inverse on the other side of the equation is to use the logic that states that one can do the same thing to both sides of an equation. As an example, one can add the same amount to both sides of an equation, subtract the same amount from both sides of an equation, multiply both sides of an equation by the same number and so on. We refer to schemes that use this kind of logic as *standard semantics schemes*. As the name implies, standard semantics with intermediate simplifications allows for the simplification of the equations that result from taking logical steps. As before, simplification begins before we take the first logical step and is performed after every logical step including the last.

To show how standard semantics with intermediate simplifications can be used to solve the problem posed above, we switch sides in the model for the problem to get

$$r \times 10 \ + \ 15.10 \times 8 \ = \ 245.80$$

We begin by simplifying the equation. This yields

$$r \times 10 \ + \ 120.80 \ = \ 245.80$$

---

[15]We will not present the case for standard semantics without intermediate simplifications as the approach is extremely inefficient and is, therefore, seldom used in practice.

Next, following the algorithm above we solve for the term that contains the unknown. This is the term $r \times 8$ and to isolate it on the left side of the equation we will have to get rid of addition of 120.80 on the left side of the equation. To do so, we subtract 120.80 from both sides of the equation. This yields

$$r \times 10 + 120.80 - 120.80 = 245.80 - 120.80$$

We now simplify the equation. On the left side, addition of 120.80 cancels with subtraction of 120.80 to leave $r \times 10$.[16] On the right side, we subtract 120.80 from 245.80 to get 125.00. This yields

$$r \times 10 = 125.00$$

We now continue with the algorithm above and solve for the factor that contains the unknown. This is the factor r and, to isolate it, we need to get rid of multiplication by 10 on the left side of the equation. To do so, we divide both sides of the equation by 10. This yields

$$\frac{r \times 10}{10} = \frac{125.00}{10}$$

We now simplify the equation. On the left side, multiplication by 10 cancels with division by 10 to leave r.[17] On the right side, we divide 125.00 by 10 to get 12.50. This yields

$$r = \$12.50/\text{h}$$

The full solution without the intervening comments is given below.

An Algebraic Solution using Standard Semantics with Intermediate Simplifications

r rate of pay at first job (\$/h)

$$245.80 = r \times 10 + 15.10 \times 8$$
$$r \times 10 + 15.10 \times 8 = 245.80$$
$$r \times 10 + 120.80 = 245.80$$

---

[16]By the statement *addition by* 120.80 *cancels out with subtraction by* 120.80 we mean that the combination turns into addition of 0 which can be dropped.

[17]By the statement *multiplication by* 10 *cancels out with division by* 10 we mean that the combination turns into multiplication by 1 which can be dropped.

$$r \times 10 + 120.80 - 120.80 = 245.80 - 120.80$$
$$r \times 10 = 125.00$$
$$\frac{r \times 10}{10} = \frac{125.00}{10}$$
$$r = \$12.50/h$$

My rate of pay at the first job is $12.50/h.

The alternative logic used in standard semantics suffers from the point of view of applicability as it is *not* in line with the manner in which we naturally reason in dealing with everyday life problems, e.g., one would *not* normally argue that *if the sum of the incomes from the two jobs equals to total income, then to find the income from the first job we subtract the income from the second job from both sides of the equation*[18] which maps onto the standard semantic move

$$r \times 10 + 120.80 = 245.80$$
$$r \times 10 + 120.80 - 120.80 = 245.80 - 120.80$$

The corresponding natural semantic move, however, follows exactly the reasoning that we typically use in dealing with everyday life problems: That *if the sum of the incomes from the two jobs equals to total income, then the income from the first job equals to total income minus the income from the second job*, i.e.,

$$r \times 10 + 120.80 = 245.80$$
$$r \times 10 = 245.80 - 120.80$$

The semantic disadvantage of standard semantics is troubling as its adoption forces one to think differently when using algebra to solve problems than when one employs logic in dealing with everyday life problems. The approach also suffers from the point of view of efficiency as it requires more writing on the left side of the equation as the solution proceeds. Their extension to rearranging formulas compounds this inefficiency even further.

Standard semantics is not recommended for use in working with algebraic equations and will no longer be used in this textbook.

Semantic vs Technical Formulations

The models that we have presented so far in the solutions to the problem above have been formulated following semantics. Semantic formulations follow the conventions of the grammar of the language: Elements of the language

---

[18]What equation??

are arranged in a certain order and the unknown is allowed to fall wherever it may. As an example of this, consider the model for the problem posed above, reproduced below

$$r \times 10 \; + \; 15.10 \times 8 \; = \; 245.80$$

By convention, within each term on the left, the rate is written first followed by time worked. But why such conventions?

The reason is that such conventions have many advantages. For one thing, preserving a certain order allows for good communication protocol. In addition, as seen earlier in the coverage of direct proportion problems, the particular order given reads better and maps well onto the general equation for direct proportion. Such order can also aid us in the task of modelling problems. As an example, as seen in the chapter on direct proportion in working with conversion of units, the arrangement of the conversion factors in the model can aid in solving problems in which the solution is modelled as a subexpression consisting of a contiguous number of such factors.

Proper order can also provide a fair amount of insight into whether or not the corresponding step-by-step solution divides into subsections, with each finding a certain value, and the values thus obtained operated on to arrive at a solution to the problem. This topic is somewhat advanced and is discussed in one of the exercises at the end of this chapter.

What semantic formulations lack, however, is that they do not readily show similarities between the *structures* of the models that one arrives at. As an example, consider the following four related models below with each having a different unknown:

$$r \times 10 \; + \; 15.10 \times 8 \; = \; 245.80$$
$$12.50 \times t \; + \; 15.10 \times 8 \; = \; 245.80$$
$$12.50 \times 10 \; + \; r \times 8 \; = \; 245.80$$
$$12.50 \times 10 \; + \; 15.10 \times t \; = \; 245.80$$

The similarity between these four equation is that, in each one, the unknown appears as a factor within a term. Indeed, the steps that go into the solution of the four models listed above are similar.

To better show such similarities, a set of conventions have been put in place which allow one to classify the model that one arrives at with ease.[19] Formulations that follow these conventions are called *technical formulations*. Technical formulations are discussed in detail in the companion textbook *Style in Technical Math* and we recommend that the reader read the relevant sections in the companion textbook before comtinuing further. Here we

---

[19]Such classifications make it easy to relate the values of the quantities involved in the model and help one deploy the proper algorithms to solve the models.

briefly note that such formulations require that, within a term, factors that involve numerical values are written on the left side of factors that involve both numerical values and unknowns and these, in turn, are written on the left side of factors that consists of unknowns only. As an example, we write $3 \times (x + 2) \times y$ in the order that they are written. Furthermore, the multiplication notation between numerical values and unknowns, as well as those that relate to brackets, are dropped. As an example, we write $8r$ and not $8 \times r$ and we write $3(x + 2)y$ in place of $3 \times (x + 2) \times y$.

It should be noted, however, that many of the advantages of semantic formulations are lost when they are paraphrased into technical formulations. For this reason, in what follows we will use semantic formulations in solutions that employ natural semantics without intermediate simplifications to preserve the full force of the algorithm. When solving a problem using natural semantics with intermediate simplifications, however, we switch to the use of technical formulations as intermediate simplifications already affect the semantics behind the solution to a great extent. The use of technical formulations may, therefore, be seen as further complicating a problem that already exists.

Here is the solution to the problem above following natural semantics with intermediate simplifications with the model re-written using a technical formulation.

An Algebraic Solution using Natural Semantics with Intermediate Simplifications using a Technical Formulation

r rate of pay at first job (\$/h)

$245.80 = 10r + 15.10 \times 8$

$10r + 15.10 \times 8 = 245.80$

$10r + 120.80 = 245.80$

$10r = 245.80 - 120.80$

$10r = 125.00$

$r = \dfrac{125.00}{10}$

$r = \$12.50/\text{h}$

My rate of pay at the first job is \$12.50/h.

In the rest of this section, we will present algebraic solutions to a number of problems that were modelled in the earlier sections. The solutions follow the algorithm introduced earlier in this section with the steps implemented using natural semantics without and with intermediate simplifications. For the first flavour we will use a semantic formulation and for the second flavour we will use a technical formulation.

Problem

The mass of a molecule of $X_6H_{12}O_6$ is 180.156 amu. Identify the unknown atom, X, by finding its atomic mass. The atomic masses of H and O are 1.008 amu/atom and 16.00 amu/atom respectively.

Here is an algebraic solution following the semantic scheme without intermediate simplifications using a semantic formulation. The model is generated by arguing that the mass of a molecule of $X_6H_{12}O_6$ is equal to the sum of the masses of X, H and O atoms in it.

An Algebraic Solution using Natural Semantics without Intermediate Simplifications using a Semantic Formulation

M atomic mass of X (amu/atom)

$$180.156 = M \times 6 + 1.008 \times 12 + 16.00 \times 6$$

$$M \times 6 + 1.008 \times 12 + 16.00 \times 6 = 180.156$$

$$M \times 6 = 180.156 - 1.008 \times 12 - 16.00 \times 6$$

$$M = \frac{108.156 - 1.008 \times 12 - 16.00 \times 6}{6}$$

$$M = \frac{180.156 - 12.096 - 96}{6}$$

$$M = \frac{72.06}{6}$$

$$M = 12.01 \text{ amu/atom}$$

The atomic mass of X is 12.01 amu/atom which identifies the atom as carbon.

We now provide some explanation as to how the solution above proceeds from one line to the next.

The first line in the solution above, i.e.,

M atomic mass of X (amu/atom)

defines the unknown and the unit in which its value will be measured. The next line provides a semantic formulation of the model. As noted above, the model states that the mass of a molecule of $X_6H_{12}O_6$ is equal to the sum of the masses of the X, H and O atoms in it.

$$180.156 = M \times 6 + 1.008 \times 12 + 16.00 \times 6$$

We switch sides to bring the unknown to the left.

$$M \times 6 + 1.008 \times 12 + 16.00 \times 6 = 180.156$$

We now solve for the term that contains the unknown.

$$M \times 6 = 180.156 - 1.008 \times 12 - 16.00 \times 6$$

We now solve for the factor that contains the unknown.

$$M = \frac{180.156 - 1.008 \times 12 - 16.00 \times 6}{6}$$

Isolation of the unknown signals the end of the logical stage. This is the beginning of the computational stage. Since the main operation on the right side is division, we need to work out the dividend and the divisor before we can carry out the division. The dividened analyzes into three terms. We work out the terms to get

$$M = \frac{180.156 - 12.096 - 96}{6}$$

We now work out the dividened

$$M = \frac{72.06}{6}$$

followed by the quotient

$$M = 12.01 \text{ amu/atom}$$

When subtracting a string of values, it is often the case that one finds the sum of these values and subtracts the total instead. As an example, if I start with \$54.00 and buy a pen for \$1.20 and a notebook for \$6.99, then I can calculate the amount of money that have left by subtracting \$1.20 from \$54.00 and then subtracting \$6.99 from the result, or, alternatively, I can add \$1.20 and \$6.99 to find the total amount of money that I sepnt and then subtract this sum from \$54.00. The second approach is often preferred as it involves fewer subtractions (especially if the list of the things that I bought is long).

Following this other line of thought, our solution would look like the following.

A Modified Algebraic Solution using Natural Semantics without Intermediate Simplifications using a Semantic Formulation

M atomic mass of X (amu/atom)

$180.156 = M \times 6 + 1.008 \times 12 + 16.00 \times 6$

$M \times 6 + 1.008 \times 12 + 16.00 \times 6 = 180.156$

$M \times 6 + (1.008 \times 12 + 16.00 \times 6) = 180.156$

$M \times 6 = 180.156 - (1.008 \times 12 + 16.00 \times 6)$

$M = \dfrac{180.156 - (1.008 \times 12 + 16.00 \times 6)}{6}$

$M = \dfrac{180.156 - (12.096 + 96)}{6}$

$M = \dfrac{180.156 - 108.96}{6}$

$M = \dfrac{72.06}{6}$

$M = 12.01$ amu/atom

The atomic mass of X is 12.01 amu/atom which identifies the atom as carbon.

Here is an algebraic solution using natural semantics with intermediate simplifications using a technical formulation.

An Algebraic Solution using Natural Semantics with Intermediate Simplifications using a Technical Formulation

M atomic mass of X (amu/atom)

$180.156 = 6M + 1.008 \times 12 + 16.00 \times 6$

$6M + 1.008 \times 12 + 16.00 \times 6 = 180.156$

$6M + 12.096 + 96 = 180.156$

$6M + 108.096 = 180.156$

$6M = 180.156 - 108.096$

$6M = 72.06$

$M = \dfrac{72.06}{6}$

$M = 12.01$ amu/atom

The atomic mass of X is 12.01 amu/atom which identifies the atom as carbon.

We let the reader extract the steps in a step-by-step solution from the alternative solutions given above.[20]

Here is a problem whose model involves brackets.

Problem

The mass of 2.8 mol of CX_2 is 123.228 g. Identify the unknown atom, X, by finding its molar mass. The molar mass of C is 12.01 g/mol.

Here is an algebraic solution following the semantic scheme without intermediate simplifications using a semantic formulation.

An Algebraic Solution using Natural Semantics without Intermediate Simplifications using a Semantic Formulation

M molar mass of X (g/mol)

$$123.228 = (12.01 + M \times 2) \times 2.8$$

$$(12.01 + M \times 2) \times 2.8 = 123.228$$

$$12.01 + M \times 2 = \frac{123.228}{2.8}$$

$$M \times 2 = \frac{123.228}{2.8} - 12.01$$

$$M = \frac{\frac{123.228}{2.8} - 12.01}{2}$$

$$M = \frac{44.01 - 12.01}{2}$$

$$M = \frac{32}{2}$$

$$M = 16 \text{ g/mol}$$

The molar mass of X is 16 g/mol which identifies the atom as oxygen.

We will now show how a set of steps in a step-by-step solution may be extracted from the solution above. We begin with the equation at the start

---

[20]You should do this. The exercise will give you a feel for one reason why it is better to postpone simplification until the unknown is solved for.

of the computational stage, i.e.,

$$M = \frac{\frac{123.228}{2.8} - 12.01}{2}$$

We begin by noting that the expression of the right side of the equation involves three operations. Our step-by-step solution will, therefore, consist of three steps.

To work out the steps, we follow the manner in which this expression is evaluated: To evaluate the expression, we would have to begin by working out the subexpression $\frac{123.228}{2.8}$. A bit of reflection shows that this aims to find the mass of 1 mol of CX_2 as it involves the division of the mass of 2.8 mol of CX_2 by 2.8. This computational step corresponds to the first step in our step-by-step solution. Following this computational step, we arrive at

$$M = \frac{44.01 - 12.01}{2}$$

We now work out the numerator $44.01 - 12.01$. A bit of reflection shows that this calculates the mass of the X atoms as the difference between the total mass and the mass of the C atoms. This computational step corresponds to the next step in our step-by-step solution. Following this computational step, we arrive at

$$M = \frac{32}{2}$$

Our last computational step involves the division of 32 by 2. This calculates the mass of 1 mol of X as it involves the division of the mass of 2 mol of X by 2. This computational step corresponds to the last step in our step-by-step solution. Following this computational step, we arrive at

$$M = 16 \text{ g/mol}$$

We can now identify the atom as oxygen.

Here is the complete step-by-step solution without the intervening comments.

A Step-by-Step Solution

1. Calculate the mass of 1 mol of CX_2.

   $$123.228 \div 2.8 = 44.01$$

   The mass of 1 mol of CX_2 is 44.01 g.

2. Calculate the mass of the X atoms.

$$44.01 - 12.01 = 32$$

The X atoms have a mass of 32 g.

3. Calculate the molar mass of X.

$$32 \div 2 = 16$$

The molar mass of X is 16 g/mol.

The ability of the language of algebra to display all of this on a single line is quite impressive:[21]

$$\frac{\frac{123.228}{2.8} - 12.01}{2}$$

It is also reassuring that one can have such confidence in the validity of the step-by-step solutions that are generated by algebraic solutions.

Here is a problem involving averages.

Problem

You wrote three tests of equal weight. You scored 75% on Test 1 and 81% on Test 2. What score on Test 3 will give you an average of 80%? The tests have equal weights.

Here is an algebraic solution following the semantic scheme without intermediate simplifications using a semantic formulation.

An Algebraic Solution using Natural Semantics without Intermediate Simplifications using a Semantic Formulation

$$g \quad \text{grade on Test 3 (1)}$$

$$80\% = \frac{70\% + 80\% + g}{3}$$

$$\frac{70\% + 80\% + g}{3} = 80\%$$

---

[21]Note that the full set of steps is on display.

$$70\% + 80\% + g = 3 \times 80\%$$
$$(70\% + 80\%) + g = 3 \times 80\%$$
$$g = 3 \times 80\% - (70\% + 80\%)$$
$$g = 240\% - 150\%$$
$$g = 90\%$$

You need a grade of 90% on Test 3 for an average of 80%.

We leave it to the reader to extract the steps for a step-by-step solution from the algebraic solution.[22] We also let the reader provide a solution using natural semantics with intermediate simplifications using a technical formulation.

We now present a problem involving tests of unequal weights.

Problem

Tests 1, 2 and 3 are worth 10%, 5% and 20% of the final mark respectively. Your scored 80% on Test 1 and 60% on Test 2. What grade on Test 3 will give you an average of 75%?

Here is an algebraic solution following the semantic scheme without intermediate simplifications using a semantic formulation.

An Algebraic Solution using Natural Semantics without Intermediate Simplifications using a Semantic Formulation

g grade needed on Test 3 for an average of 75% (1)

$$75\% = \frac{0.1 \times 80\% + 0.05 \times 60\% + 0.2 \times g}{0.1 + 0.05 + 0.2}$$

$$\frac{0.1 \times 80\% + 0.05 \times 60\% + 0.2 \times g}{0.1 + 0.05 + 0.2} = 75\%$$

---

[22]Variations in step-by-step solutions can always be expressed in algebraic forms. The steps that we promote in solving equations in this textbook assure that the corresponding logic is both in line with the manner in which we naturally reason and, of the possible variations in doing so, it follows the logic that is most efficient.

$$0.1\,(80\%) + 0.05\,(60\%) + 0.2\,(g) = (0.1 + 0.05 + 0.2)\,(75\%)$$

$$0.2\,(g) = (0.1 + 0.05 + 0.2)\,(75\%) - 0.1\,(80\%) - 0.05\,(60\%)$$

$$g = \frac{(0.1 + 0.05 + 0.2)\,(75\%) - 0.1\,(80\%) - 0.05\,(60\%)}{0.2}$$

$$g = \frac{0.35 \times 75\% - 8\% - 3\%}{0.2}$$

$$g = \frac{26.25\% - 8\% - 3\%}{0.2}$$

$$g = \frac{15.25\%}{0.2}$$

$$g = 76.25\%$$

I need 76.25% on Test 3 for an average of 75%.

It is somewhat challenging to turn this solution into a step-by-step solution but we do invite the reader to give it a try.

Here is the alternative using natural semantics with intermediate simplifications using a technical formulation.

An Algebraic Solution using Natural Semantics with Intermediate Simplifications using a Technical Formulation

g grade needed on Test 3 for an average of 75% (1)

$$75\% = \frac{0.1 \times 80\% + 0.05 \times 60\% + 0.2g}{0.1 + 0.05 + 0.2}$$

$$\frac{0.1 \times 80\% + 0.05 \times 60\% + 0.2g}{0.1 + 0.05 + 0.2} = 75\%$$

$$\frac{8\% + 3\% + 0.2g}{0.35} = 75\%$$

$$\frac{11\% + 0.2g}{0.35} = 75\%$$

$$11\% + 0.2g = 0.35 \times 75\%$$

$$11\% + 0.2g = 26.25\%$$

$$0.2g = 26.25\% - 11\%$$

$$0.2g = 15.25\%$$

$$g = \frac{15.25\%}{0.2}$$

$$g = 76.25\%$$

I need 76.25% on Test 3 for an average of 75%.

Here is a problem involving discount and tax.

Problem

A hat, priced at \$15.20 was on sale. Tax was added at 13% of the sale price. Find the rate of discount if the grand total comes to \$14.60.

Here is an algebraic solution using natural semantics without intermediate simplifications using a semantic formulation.

An Algebraic Solution using Natural Semantics without Intermediate Simplifications using a Semantic Formulation

$$r \quad \text{rate of discount (1)}$$
$$14.60 = (1 + 0.13) \times (1 - r) \times 15.20$$
$$(1 + 0.13) \times (1 - r) \times 15.20 = 14.60$$
$$1 - r = \frac{14.60}{(1 + 0.13) \times 15.20}$$
$$-r = -1 + \frac{14.60}{(1 + 0.13) \times 15.20}$$
$$r = 1 - \frac{14.60}{(1 + 0.13) \times 15.20}$$
$$r = 1 - \frac{14.60}{1.13 \times 15.20}$$
$$r = 1 - 0.850\,023\ldots$$
$$r = 0.149\,976\ldots$$

The rate of discount is 15%.

Once again we challenge the reader to work out a step-by-step solution from the algebraic solution given above.[23]

We present an algebraic solution using natural semantics with intermediate simplifications using a technical formulation.

---

[23]See the exercises at the end of this section.

An Algebraic Solution using Natural Semantics with Intermediate Simplifications using a Technical Formulation

r rate of discount (1)

$14.60 = (1 + 0.13)(15.20)(1 - r)$

$(1 + 0.13)(15.20)(1 - r) = 14.60$

$1.13(15.20)(1 - r) = 14.60$

$17.176(1 - r) = 14.60$

$1 - r = \dfrac{14.60}{17.176}$

$1 - r = 0.850\,0232\ldots$

$-r = 0.850\,023\ldots - 1$

$-r = -0.149\,976\ldots$

$r = 0.149\,976\ldots$

The rate of discount in 15%.

It is instructive to try to extract a step-by-step solution from the algebraic solution above as the simplification resulting in the value 17.176 disrupts the natural flow of logic.[24]

Problems that deal with percentages often exhibit linear behaviour with respect to their unknown. Here is an example.

Problem

16.5% of the frogs in the sample have red eyes. How many frogs are there in the sample if 1048 frogs in the sample do *not* have red eyes?

Here is an algebraic solution using natural semantics without intermediate simplifications using a semantic formulation.

---

[24] See the exercises at the end of this section.

An Algebraic Solution using Natural Semantics without Intermediate Simplifications using a Semantic Formulation

n number of frogs in the sample (1)

$1048 = (1 - 0.165) \times n$

$(1 - 0.165) \times n = 1048$

$n = \dfrac{1048}{1 - 0.165}$

$n = \dfrac{1048}{0.835}$

$n = 1255.089 \ldots$

There are 1255 frogs in the sample.

Here is an algebraic solution using natural semantics with intermediate simplifications using a technical formulation.

An Algebraic Solution using Natural Semantics with Intermediate Simplifications using a Technical Formulation

n number of frogs in the sample (1)

$1048 = (1 - 0.165)\,n$

$(1 - 0.165)\,n = 1048$

$0.835n = 1048$

$n = \dfrac{1048}{0.835}$

$n = 1255.089 \ldots$

There are 1255 frogs in the sample.

Exercise Set 9.3A

For each problem in Exercise Set 9.2, solve the given model for the unknown using

1. natural semantics without intermediate simplifications using a semantic formulation.
2. standard semantics with intermediate simplifications using a technical formulation.

Exercise Set 9.3B

1. Consider the following problem.

> You wrote three tests of equal weight. You scored 75% on Test 1 and 81% on Test 2. What score on Test 3 will give you an average of 80%? The tests have equal weights.

Here is an algebraic solution following the semantic scheme without intermediate simplifications using a semantic formulation.

$$g \quad \text{grade on Test 3 for the desired average (1)}$$

$$80\% = \frac{75\% + 81\% + g}{3}$$

$$\frac{75\% + 81\% + g}{3} = 80\%$$

$$75\% + 81\% + g = 3 \times 80\%$$

$$(75\% + 81\%) + g = 3 \times 80\%$$

$$g = 3 \times 80\% - (75\% + 81\%)$$

$$g = 240\% - 156\%$$

$$g = 84\%$$

You need a grade of 84% on Test 3 for an average of 80%.

Convert this to a step-by-step solution.

2. Consider the following problem.

> 16.5% of the frogs in the sample have red eyes. How many frogs are there in the sample if 1048 frogs in the sample do *not* have red eyes?

An algebraic solution using natural semantics without intermediate simplifications using a semantic formulation is given below.

$$n \quad \text{number of frogs in the sample (1)}$$

$$1048 = (1 - 0.165) \times n$$

$$(1 - 0.165) \times n = 1048$$

$$n = \frac{1048}{1 - 0.165}$$

$$n = \frac{1048}{0.835}$$

$$n = 1255.089 \ldots$$

There are 1255 frogs in the sample.

Convert this to a step-by-step solution.

3. Consider the following problem.

> 40% of the frogs in the sample have red eyes. Of those that have red eyes, 90% have green skin. Of the red-eyed frogs that have green skin, some grow into adult frogs. Of the red-eyed frogs that have green skin and grow into adult frogs, 80% will have offspring. What percentage of the red-eyed, green-skinned frogs grow into adult frogs if 302 red-eyed, green-skinned frogs are expected to grow into adult frogs and have offspring? There are 1500 frogs in the sample.

An algebraic solution using natural semantics without intermediate simplifications using a semantic formulation is given below.

n number of red-eyed, green-skinned, adult frogs in the sample that are expected to have offspring (1)

$$302 = 0.8 \times r \times 0.9 \times 0.4 \times 1500$$

$$0.8 \times r \times 0.9 \times 0.4 \times 1500 = 302$$

$$r = \frac{302}{0.8 \times 0.9 \times 0.4 \times 1500}$$

$$r = 0.699\ldots$$

70% of red-eyed, green-skinned frogs grow into adult frogs.

Convert this to a step-by-step solution.

4. Consider the following problem.

> A hat, priced at $15.20 was on sale. Tax was added at 13% of the sale price. Find the rate of discount if the grand total came to $14.60.

An algebraic solution using natural semantics without intermediate simplifications using a semantic formulation is given below.

r rate of discount (1)

$$14.60 = (1 + 0.13)(1 - r)(15.20)$$

$$(1 + 0.13)(1 - r)(15.20) = 14.60$$

$$1 - r = \frac{14.60}{(1 + 0.13)(15.20)}$$

$$-r = -1 + \frac{14.60}{(1 + 0.13)(15.20)}$$

$$r = 1 - \frac{14.60}{(1 + 0.13)(15.20)}$$

$$r = 1 - \frac{14.60}{1.13 \times 15.20}$$

$$r = 1 - 0.850\,023\ldots$$

$$r = 0.149\,976\ldots$$

The rate of discount is 15%.

Convert this to a step-by-step solution.

Chapter 10
Linear Equations in Two Unknowns

An equation in two unknowns[1] can be used to evaluate the value of one of the unknowns given the value of the other unknown. In addition, they provide us with visual cues that allow us to relate the value of one unknown to the value of the other unknown. This classification, in turn, helps us make conclusions as to the manner in which changes in the value of one unknown affects the value of the other unknown.

Linear equations in two unknowns are equations whose form maps onto the general linear equation $y = mx + b$ where m and b are constants that represent the slope and y-coordinate of the y-intercept of the associated graph which is a straight line.[2]

In this chapter we will discuss some of the semantic features of linear equations in two unknowns. Our focus is mainly on the interpretation of the slope and intercepts of the linear equation in the context of the word problems that we explore.

Conversion of Units of Temperature

Consider the following formula for the conversion of the unit of temperature from degrees Celsius to kelvins.

T temperature (K)
t temperature ($^{\circ}$C)
T_0 temperature at which water freezes under
         standard conditions (K)

$$T = T_0 + t$$

---

[1] In applications, each unknown represents the value of a quantity.

[2] Technical aspects of linear equations including their graphs and the concepts of slope and intercepts are discussed at length in the companion textbook *Style in Technical Math* by the author and we invite the reader to read this material before continuing further. The discussion that follows assumes this background on the part of the reader.

In this equation T_0 is a constant with a defined value of 273.15 K.

Let us analyze this equation in detail.

Consider the Celsius scale of temperature displayed in Figure 10.0.1 below.[3]

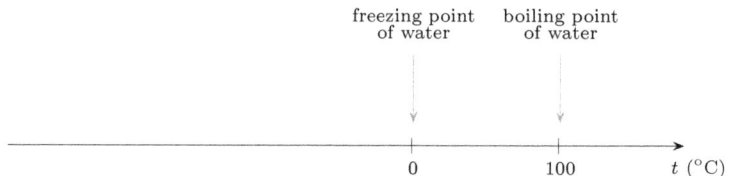

freezing point   boiling point
of water   of water

0   100 t (°C)

Figure 10.0.1: The Celsius scale

The definition of this unit of temperature is such that it assigns the value 0 to the freezing point of water and 100 to the boiling point of water, thus generating 100 divisions from the freezing point of water to the boiling point of water. This is a very useful scale of temperature for use by humans as water is a substance that we work with frequently and its freezing and boiling points are of importance to us. However, the scale is not a useful one when working with scientific theories that involve temperature as temperature is a general concept and, therefore, neither the definition of the concept nor the design of the unit to measure its value with should be based on the behaviour of a specific substance.

The temperature of an object is a measure of the speed with which its molecules move around and about. This speed, in turn, relates to the kinetic energy of the molecules in the object. The sum of the kinetic energies of all the molecules in an object is called the heat content of the object.[4] The higher the

---

[3]The freezing and boiling points of water are assumed to occur under standard conditions.

[4]The actual picture is more complex: The heat content of an object is defined as the sum of the various forms of energies of the molecules within that object. These include translational energy of the molecules, rotational energy of the molecules, vibrational energy of the molecules, and any energies that are associated with the movement of the electrons within the atoms of these molecules.

When heat is added to an object, the heat content of the object increases and this increase appears as an increase in various forms of energy that are listed above. A portion of the heat, as an example, might increase the translational energies of the molecules and another portion might increase the rotational energies of the molecules.

Temperature, in its ordinary sense, is a measure of only the translational energies of the molecules within a substance. What this means is that, an increase in temperature in its ordinary sense measures the increase in the amount of only the translational energies of the molecules in that substance. If, as an example, the heat that enters a substance is used to increase the rotational energies of the molecules in that substance, then the temperature of the substance in its ordinary sense will *not* rise even though heat is entering the system

temperature, the faster the molecules within an object move and the higher the kinetic energies of the molecules. The sum of these energies is referred to as the heat content of the object. Since temperature is a measure of the speed of the molecules in a given substance, it makes sense to define a unit of temperature that sets 0 as the value at which the molecules within a substance stop moving.[6] As you can see, this definition applies to all substances and does not rely on the properties of a specific substance like water.

Our new scale of temperature so far is shown in Figure 10.0.2 below next to the Celsius scale.

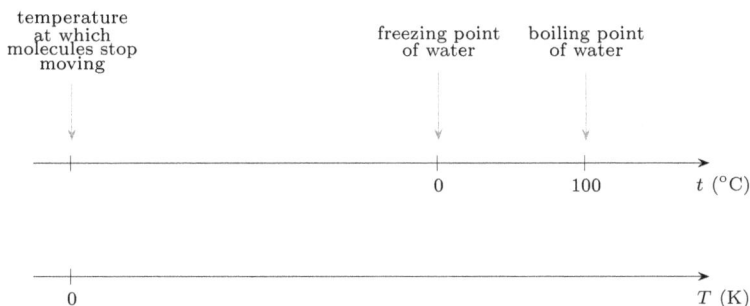

Figure 10.0.2: The Celsius scale versus to the kelvin scale

(and, therefore, the heat added maybe thought of as a sort of *latent heat* – See below.)

Under ordinary conditions, the portion of heat that affects the rotational, vibrational, and electronic energies of the molecules within a substance is quite small compared to the portion that affects the translational energies of the molecules of the substance. For this reason, it is common to assume that added heat increases the temperature in its ordinary sense unless of course the heat is used to either break up intermolecular bonds such as hydrogen bonds in water, atomic bonds or in ionizing the atoms within the substance in which case the temperature of the object in its ordinary sense remains constant until these processes complete (e.g., until all hydrogen bonds are broken) at which time any additional heat will return to affecting the translational energies of the molecules and temperature, in its ordinary sense, rises.[5]

Since temperature in its ordinary sense is a measure of the kinetic energies of the molecules of a substance only, some have chosen to call it the *translational temperature* to differentiate between this temperature, which is temperature in its ordinary sense, and other temperatures that are used to measure the rotational, vibrational and electronic energies of the molecules, atoms and electrons within a substance.

In the discussion that follows in the body of the textbook, we will use the word temperature in its ordinary sense and assume that any heat added or removed from the system will affect only the translational energies of the molecules within that system.

[6]This association of the 0s, i.e., 0 K and 0 speed of molecules, is important as it implies that any quantity whose value is proportional to heat will be proportional to temperature are well. This explains why, in the Ideal Gas Equation of State, $pV = nRT$, we prefer to have the temperature in kelvins as this implies a proportion relationship between the temperature and pressure, volume and amount. The use of degrees Celsius generates nontrivial linear relationships between temperature and pressure, volume and amount as the formula would have to be written as $pV = nR(T_0 + t)$.

While the new scale reassigns 0 on the Celsius scale, we still prefer to keep 100 divisions between the freezing and boiling points of water as this makes it easier to work with the range of temperatures that we normally deal with. It also makes it easier to relate the two units of temperature, kelvins and degrees Celsius. With this added requirement, our new scale of temperature is shown in Figure 10.0.3 below.

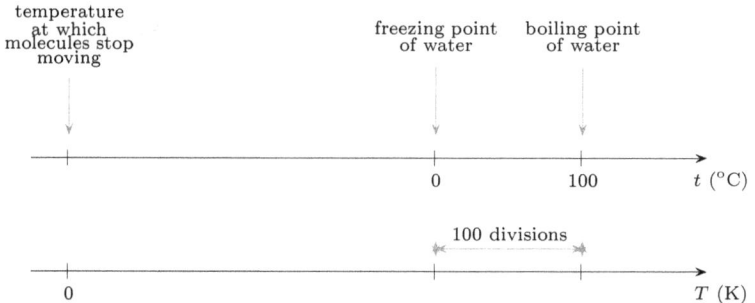

Figure 10.0.3: The Celsius scale versus to the kelvin scale

Note that the requirement to divide the range from the freezing point of water to the boiling point of water into 100 parts effectively sets the size of the unit on the new scale. Furthermore, since the same number of divisions between the freezing and boiling points of water are used by the Celsius scale, this implies that the size of the unit is the same on both scales, i.e., a change of 1 °C implies a change of 1 K and vice versa. Following this size down from the freezing point of water to the point where molecules stop moving, we find that we need 273.15 such steps. Retaking our steps back up from the point where molecules stop moving, which is now set to be 0 on the new scale, yields the value 273.15 K as the freezing point of water and 373.15 K as the boiling point of water. We now formally define T_0 as 273.15 K.

We refer to the new scale as the kelvin scale with unit symbol K. The kelvin scale is shown in Figure **??** below. Superimposed on this diagram is an example of how the numerical value of T, the value of the temperature in kelvins, maps onto the sum of 273.15 and the numerical value of t, the value of the temperature in degrees Celsius.

The formula relating the two scales of temperature, i.e.,

$$T = T_0 + t$$

then, implies that, to find the temperature in kelvins, we can start with T_0, i.e.,

273.15 K

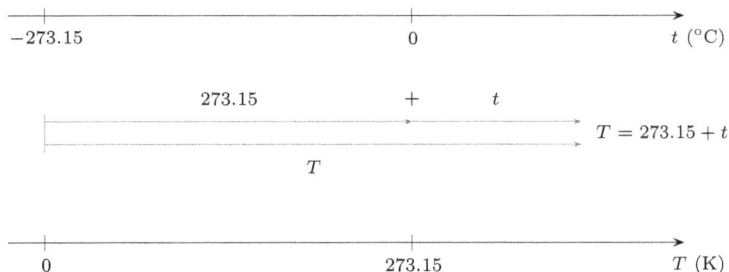

Figure 10.0.4: Relationship between the Celsius and kelvin sclaes

and then, for each degree Celsius, we add one K, i.e.,

$$\frac{1 \text{ K}}{1 \text{ }^\circ\text{C}} \times t \; + \; 273.15$$

Note how the unit °C within the first term cancels out, leaving the unit K as the unit of the term.

The equation

$$T \; = \; T_0 \; + \; t$$

has the form of the linear equation, $y = mx + b$ with m, the slope of the associated line, equal to 1 K/°C, and a T-intercept of $(0, T_0)$, i.e., $(0, 273.15)$. The t-intercept can be found by setting T equal to 0 in the formula above and solving the resulting equation for t.

$$
\begin{aligned}
T &= T_0 + t \\
T_0 + t &= T \\
t &= T - T_0 \\
t &= 0 - T_0 \\
t &= -T_0
\end{aligned}
$$

which leads to $(-T_0, 0)$ or $(-273.15, 0)$ as the t-intercept.

The graph of the formula $T = T_0 + t$ is given in Figure 10.0.5 below.[7]

Two of the key concepts that need close attention in linear relationships are the concepts of slope and intercepts of the graph of the associated equation. The reason these concepts are important is that often their values have significance in the context of the word problem.

---

[7]The presence of the heavy dot on the left side of the line in the figure above points to the coldest temperature. This corresponds to 0 K and −273.15 °C and represents the temperature at which molecules stop moving.

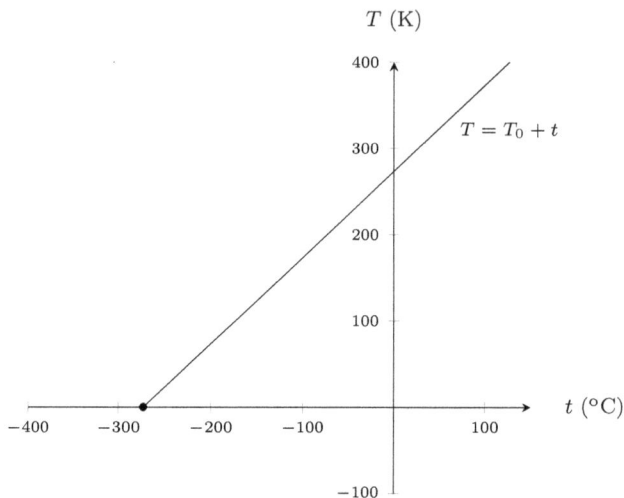

Figure 10.0.5: Graph of the equation $T = T_0 + t$

As an example, consider the slope, m, of the graph for the formula for the conversion of the unit of temperature from degrees Celsius to kelvins given above. The slope of this line is 1 K/°C. This implies that if the temperature changes by 1 °C, then it has changed by 1 K which in turn implies that the size of the two units of temperature are the same.

The t-intercept of this line is $(-T_0, 0)$. The significance of this intercept is that it provides us with the value of the temperature at which molecules stop moving in both kelvins and degrees Celsius, i.e., 0 K and -273.15 °C.

The T-intercept of this line is $(0, T_0)$. The significance of this intercept is that it provides us with the value of the temperature at which water freezes in both kelvins and degrees Celsius, i.e., 273.15 K and 0 °C.

Conversion of the unit of temperature from degrees Celsius to degrees Fahrenheit is left as an exercise.

Motion along a Straight Line at Constant Speed

The description of change in physics often begins by a study of *kinematics* which is the branch of physics that deals with the description of motion as opposed to its cause.[8] A problem that is commonly seen in the study of kinematics is the problem of describing the motion of an object moving

---

[8]The latter is studied under the branch of physics called *dynamics*.

along a straight line at constant speed. By *describing the motion* we mean identifying the location of the moving object as time goes on.

Here is an example of such a problem.

Problem

An object is initially 50 km from an observer, moving away from the observer along a straight line at a speed of 100 km/h. Find a relationship between final distance between the object and the observer and time passed.

In general, for such a problem, the final distance between the object and the observer, which we will refer to as d, is set equal to the sum of the initial distance between the object and the observer, which we refer to as d_0, and the distance that the object covers as time passes. The latter distance may be computed by multiplying the speed of the object, which we represent as v, and time passed, which we represent as t. This yields the following.

d final distance between object and observer (km)
d_0 initial distance between object and observer (km)
v speed of the object (km/h)
t time passed (h)

$$d = d_0 + vt$$

The problem posed earlier is a special case of this more general problem where the initial distance, d_0, is given as 50 km and the speed of the object, v, is given as 100 km/h. This turns the equation above into

d final distance between object and observer (km)
t time passed (h)

$$d = 100t + 50$$

This is a linear equation in two unknowns as it has the general form of the linear equation, $y = mx + b$, with m equal to 100 km/h and b equal to 50 km. The graph of this equation is given in Figure 10.0.6 below.

We now look at the significance of the slope of the line and its intercepts.

The slope of the line is, as noted above, equal to 100 km/h and, therefore, represents the speed of the object. This means that if we move along the positive t direction over 1 h (i.e., if the object moves for an hour), then we will move along the d-axis over 100 km (the object covers 100 km).

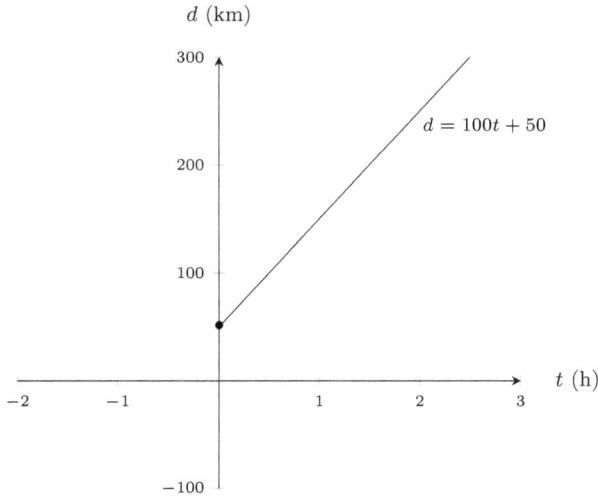

Figure 10.0.6: Graph of the equation $d = 100t + 50$

The d-intercept is $(0, 50)$ which indicates that when timing started (when $t = 0$ h), the object was 50 km from the observer ($d = 50$ km) and, therefore, represents the initial distance between the object and the observer.

The t-intercept is $(-0.5, 0)$. This implies that at t equal to -0.5 h (half an hour before timing started), the object was at location 0 (where the observer is). In the context of the present problem this may or may not be meaningful: If the object started at a distance of 50 km when timing started, then the t intercept does not have significance as it points to the location of the object half an hour ago which is not of interest. However, if the object has been in motion at the same speed prior to the observer beginning to time it, then we can say that half an hour ago ($t = -0.5$ h), the object was where the observer is ($d = 0$ km).

As the above two cases show, linear equations relate two quantities where the value of one of them (i.e., y) can be computed by adding a fixed amount (i.e., b) to a variable amount (i.e., mx) with the variable amount being directly proportional to the value of the other quantity (mx is directly proportional to x). For such problems, the actual values of the two quantities are linearly dependent on each other but changes in the values of two quantities are directly proportional to each other.

A few other cases whose models are linear are discussed in the exercises.

Exercise Set 10

1. The following formula can be used to convert the unit of temperature from degrees Celsius to degrees Fahrenheit.

   $t_{\circ C}$ temperature $\left(^{\circ}C\right)$
   $t_{\circ F}$ temperature $\left(^{\circ}F\right)$

   $$t_{\circ F} = \frac{9}{5}t_{\circ C} + 32$$

   a. Graph the equation above.
   b. Calculate or otherwise determine the slope of the line.
   c. Calculate or otherwise determine the intercepts of the line.
   d. Interpret the slope and the intercepts in the context of the word problem.

2. The following formula can be used to compute the amount of loan that has been repaid for a purchase with a down payment of $5000 and monthly payments of $400/month.

   a amount of loan repaid ($)
   t time passed (months)

   $$a = 400t + 5000$$

   a. Graph the equation above.
   b. Calculate or otherwise determine the slope of the line.
   c. Calculate or otherwise determine the intercepts of the line.
   d. Interpret the slope and the intercepts in the context of the word problem.

3. Superposition problems and direct proportion problems are special cases of linear problems. The formula for converting the unit of temperature from degrees Celsius to kelvins discussed in the body of the text is an example of a linear equation that is also direct superposition.

   The problem below is an example of a linear equation that is also direct proportion. When a linear relationship is of type direct proportion, not only are changes in the values of the quantities involved related through direct proportion, but so are the actual values of the quantities themselves.

   An example of a linear equation that is also direct proportion is the following:

   Convert the unit of length from cm to m given that 1 m = 100 cm.

The following formula models the problem.

l_m length (m)
l_cm length (cm)

$$l_\text{m} = \frac{1}{100}l_\text{cm}$$

a. Graph the equation above.
b. Calculate or otherwise determine the slope of the line.
c. Calculate or otherwise determine the intercepts of the line.
d. Interpret the slope and the intercepts in the context of the word problem.
e. How can you tell, by looking at the graph of this equation, that it represents a relationship of type direct proportion?

Part V
Nonlinearity

Nonlinear problem are problems whose solutions go beyond the determination of the value of a factor. In this part of the book we will consider nonlinear problems that involve division, exponents, roots and logarithms. We will limit the discussion to the simplest types of these operands.

Chapter 11
Nonlinear Equations in One Unknown

In this chapter we will discuss a few select problems whose models are nonlinear with the model involving a single unknown appearing only once throughout the equation.[1]

11.1   Acidic and Basic Solutions

Acidic and basic solutions owe their existence to an imbalance between the number of H^+ and OH^- ions in water.[2] Such ions are generated through the relatively infrequent self-ionization of water molecules according to the following reaction.[3]

$$H_2O \quad \rightarrow \quad H^+ \quad + \quad OH^-$$

Of course it might happen every now and then that an H^+ ion may meet an OH^- ion and two might recombine to form a water molecule according to the reaction

$$H^+ \quad + \quad OH^- \quad \rightarrow \quad H_2O$$

It is common to use a double arrow to indicate that the reaction may go

---

[1] For background material on nonlinear equation including solution algorithms we refer the reader to the companion textbook *Style in Technical Math* by the author.

[2] We refer to an H^+ ion as a **hydrogen ion** and we call an OH^- ion a **hydroxide ion**.

[3] It is now known that an H^+ ion combines with another water molecule to generate an H_3O^+ ion, called a **hydronium ion**. However, the use of H^+ has persisted and, so long as concepts surrounding the acidity and basicity of a solution is concerned, it makes no difference whether the positive ions are seen as H^+ or H_3O^+. In this textbook we will stick to the notation H^+. The reader who prefers to do so, may replace these instances of H^+ with H_3O^+.

either way.[4]

$$H_2O \quad \rightleftharpoons \quad H^+ \quad + \quad OH^-$$

Note that, in pure water, the concentrations of H^+ and OH^- ions are equal. We write this as[5]

$$[H^+] \; = \; [OH^-]$$

The reason for the equal numbers of H^+ and OH^- ions is that the forward reaction generates an equal number of these ions: one of each. And the backward reaction destroys an equal number of these ions: one of each. Since concentration relates to the number of entities per litre of a solution, equal numbers of H^+ and OH^- ions implies equal concentrations of these ions.

Based on measurements, the concentrations of H^+ and OH^- ions in pure water are equal to 1×10^{-7} mol/L. This is equal to 0.000 000 1 mol/L, i.e., one ten millionth of a mol per litre of solution. Since 1 mol of a substance is made up of 6.02×10^{23} particles of that substance, a concentration of 1×10^{-7} mol/L implies that there are

$$\frac{6.02 \times 10^{23} \text{ particles}}{1 \text{ mol}} \times \frac{1 \times 10^{-7} \text{ mol}}{1 \text{ L}}$$

or 6.02×10^{16} particles/L. There are, therefore, 6.02×10^{16} H^+ ions and 6.02×10^{16} OH^- ions per litre of solution. This is some 60 200 000 000 000 000 ions of each kind.

---

[4]The relative lengths of the arrows indicate the relative frequencies with which the forward and backward reactions occur. As an example, if the forward reaction occurs more frequently than the backward reaction, then we will use a longer arrow for the forward reaction than the backward reaction, i.e., we write

$$H_2O \quad \rightleftharpoons \quad H^+ \quad + \quad OH^-$$

and if the backward reaction occurs more frequently than the forward reaction, then we will use a longer arrow for the backward reaction than the forward reaction, i.e., we write

$$H_2O \quad \rightleftharpoons \quad H^+ \quad + \quad OH^-$$

The larger the difference between the lengths of the arrows, the larger the difference between the frequencies with which they occur.

[5]The word *concentration* is a quantity name with the associated quantity symbol, c. Concentration is a measure of the number of entities of a certain kind in a unit volume of a solution. The SI unit of concentration is mol/m^3.

Notwithstanding the conventions above on the quantity symbol and unit of concentration, it is quite common in chemistry textbooks to see alternative symbols and units: For the quantity symbol, it is common to see the use of square brackets, e.g., they write $[H^+]$ in place of c_{H^+}. And for the unit, it is common to see the unit mol/L used in place of mol/m^3.

In this textbook we will follow the common agreement to adopt the conventions of a particular field and use the quantity symbol and unit that are commonly used in chemistry textbooks.

This might sound like a huge number – and it is – but it pales in comparison to the number of H_2O molecules that are present in one litre of the solution. To see this, we will compute the number of H_2O molecules in 1 litre of the solution and compare this to the number of H^+ and OH^- ions in one litre of the solution. Given the density of water, 997 g/L, and its molar mass, 18.016 g/mol, the number of water molecules in 1 L of water can be computed as

$$\frac{6.02 \times 10^{23} \text{ molecules}}{1 \text{ mol}} \times \frac{1 \text{ mol}}{18.016 \text{ g}} \times \frac{997 \text{ g}}{1 \text{ L}} \times 1 \text{ L}$$

which simplifies to the unit rate $3.331 \ldots \times 10^{25}$ molecules of H_2O per litre of the solution. Comparing this to the number of H^+ or OH^- ions in 1 litre of the solution, we have

$$\frac{3.331 \ldots \times 10^{25} \text{ } H_2O \text{ molecules}}{6.02 \times 10^{16} \text{ } H^+ \text{ or } OH^- \text{ ions}}$$

which simplifies to the unit rate $5.533 \ldots \times 10^8$ molecules of H_2O per ion of H^+ or OH^-, i.e., there are roughly 550 million molecules of water for each H^+ or OH^- ion. Therefore, in pure water, there are extremely few H^+ or OH^- ions compared to H_2O molecules.

As we noted above, in pure water there are an equal number of H^+ and OH^- ions.[6] We can, however, upset this balance by dissolving substances in water that generate H^+ ions or OH^- ions. Substances that, when dissolved in water, generate H^+ ions are referred to as **acids** and a solution that has extra H+ ions compared to OH^- ions is referred to as an **acidic solution**. Similarly, substances that, when dissolved in water, generate OH^- ions are referred to as **bases** and a solution that has extra OH^- ions compared to H^+ ions is referred to as a **basic solution**.

An example of a substance that, when dissolved in water, generates H^+ ions is hydrochloric acid, HCl:[7]

$$HCl_{(aq)} \quad \rightarrow \quad H^+ \quad + \quad Cl^-$$

and an example of a substance that, when dissolved in water, generates OH^- ions is sodium hydroxide, NaOH:

$$NaOH_{(aq)} \quad \rightarrow \quad Na^+ \quad + \quad OH^-$$

Experimental data seem to indicate that the relationship between $[H^+]$ and $[OH^-]$ is of type inverse proportion. As an example, when the number of

---

[6]Such a solution is said to be **neutral**.

[7]The subscript aq on HCl stands for the word *aqueous* which is interpreted to mean that HCl is dissolved in water.

H^+ ions becomes 4 times smaller, the number of OH^- ions becomes 4 times larger and when the number of H^+ ions becomes 10 times larger, the number of OH^- ions becomes 10 times smaller and so on. Let us see if we can explain this behaviour.

We begin with 1 L of pure water. As we have seen before, the following forward and backward reactions take place in pure water:

$$H_2O \;\rightleftharpoons\; H^+ \;+\; OH^-$$

with the equality in the lengths of the forward and backward arrows indicating that the forward and backward reactions happen as frequently. Since the forward reaction generates a pair of H^+ and OH^- ions while the backward reaction destroys them by combining them into H_2O, the overall balance of the number of H^+ ions and OH^- ions remains the same. We refer to such a scenario as **equilibrium**.

We now upset this equilibrium by adding a substance that generates H^+ ions until the concentration of H^+ increases to 10 times its original value. Since there are 10 times as many H^+ ions, the frequency with which an H^+ ion meets an OH^- ion increases 10 times. This results in the backward reaction to speed up 10 times while the forward reaction remains unaffected, i.e., we now have the following scenario:

$$H_2O \;\rightleftharpoons\; H^+ \;+\; OH^-$$

Since the backward reaction occurs more frequently, as time proceeds, there will be a net loss of H^+ and OH^- ions. We cannot, of course, allow the concentration of H^+ to drop below 10 times its original value as it is our aim to find out what happens to the concentration of OH^- ions *if the concentration of H^+ ions increases to 10 times its original value.* To keep the concentration of H^+ equal to 10 times its original value, we will replace any H^+ ions that may be lost as time goes on. As an example, if, in a certain length of time, 1 H^+ ion is generated by the breakdown of water (the forward reaction) while 10 H^+ are lost through recombination with OH^- (the backward reaction), we will add 9 H^+ ions to the solution to compensate for the loss.

We keep adding H^+ as needed to keep its concentration at 10 times its original value. However, we will not interfere with the decrease in the concentration of OH^- as it is our aim to see how the concentration of OH^- changes in response to the jump in the concentration of H^+. As the number of OH^- ions decreases through the more frequent backward reaction, it becomes less and less likely that an H^+ ion meets an OH^- ion and the backward reaction slows down and the length of the arrow representing the backward reaction decreases.[8] The number of OH^- ions decreases until the backward reaction

---

[8]The forward reaction continues to remain unaffected.

occurs as frequently as the forward reaction. At this point we are back at equilibrium and we no longer have to compensate for losses of H^+ ions as any loss is compensated by the equally frequent forward reaction.

By the time we reach equilibrium, the backward reaction has slowed 10 times compared to the time when it speeded up 10 times due to the sudden initial increase in the number of H^+ ions and this decrease is entirely due to the drop in the number of OH^- ions as we kept the number of H^+ ions constant at 10 times its original value. Since the reaction has slowed down 10 times, and since this drop is entirely due to the drop in the number of OH^- ions, then the number of OH^- ions must have dropped 10 times as this would lower the frequency with which an H^+ ion meets an OH^- ion 10 times which in turn lowers the frequency with which the backward reaction occurs 10 times.[9]

Based on the argument above, an increase in the number of H^+ ions in pure water by a factor of 10 forces a decrease in the number of OH^- ions by a factor of $\frac{1}{10}$. The argument holds for any number, not just 10: We could go through the same argument, this time increasing the number of H^+ ions by a factor of 2 and we would arrive at the conclusion that the number of OH^- ions decreases in response by a factor of $\frac{1}{2}$, etc.[10] This shows that the relationship between the concentrations of H^+ and OH^- ions is one of inverse proportion, as shown by experimental data.[11]

The equation relating $[H^+]$ and $[OH^-]$ must, therefore, have the same form as the equation for inverse proportion problems. Either flavour may be used: The standard form, i.e., $q_2 = \frac{k}{q_1}$, or the conservation form, i.e., $q_2 q_1 = k$. In the context of acidic and basic solutions, it is common to use

---

[9]In a sense, then, we have reached a new equilibrium; one with a composition that is different from the original. Note that in the original composition there were an equal number of H^+ and OH^- ions in the solution whereas now there are 10 times as many H^+ ions and 10 times fewer OH^- ions in the solution. As an example, if the initial solution contained 1000 H^+ ions and 1000 OH^- ions, it now contains 10 000 H^+ ions and 100 OH^- ions.

[10]It is important to keep in mind that the response is not instantaneous: It takes time for the system to move from its initial state of equilibrium to its final state of equilibrium.

[11]The reader should note the interplay between experimental and theoretical science. Experimental science attempts to relates the values of quantities through the results of experiments whereas theoretical science attempts to provide an explanation for this relationship in the simplest possible manner.

The philosophy of science grants experimental results the upper hand: In case there is disagreement between the conclusions of a theoretical argument and experimental data, the theoretical argument is discarded – assuming of course that the experimental setup is sound.

Note also that it is quite possible to come up with multiple theoretical explanations to describe the results of experimental data. In such cases, the philosophy of science grants validity to the theoretical explanation that is the simplest.

the conservation form, i.e., we write

$$[H^+][OH^-] = k$$

where k is a constant. To find the value of this constant, we can use any related pair of values of $[H^+]$ and $[OH^-]$.[12] The simplest pair of values to use are those in pure water, i.e., 1×10^{-7} mol/L. This yields the following.

$$[H^+][OH^-] = k$$
$$k = [H^+][OH^-]$$
$$k = 1 \times 10^{-7} \times 1 \times 10^{-7}$$
$$k = 1 \times 10^{-7-7}$$
$$k = 1 \times 10^{-14} \text{ mol}^2/\text{L}^2$$

The representation of the numerical value is usually shortened to 10^{-14}.

The quantity represented by k is called the **dissociation constant** for water and its value is denoted using the quantity symbol K_w. With this notation, we now have the following formula.[13]

$$[H^+][OH^-] = K_w$$

where the measured value of K_w is 10^{-14} mol^2/L^2.

Typical numerical values representing the concentrations of H^+ and OH^- ions in solutions fall between 1×10^{-1} and 1×10^{-14}. This means that it is common to see such numerical values as 3.2×10^{-4} and the like for the concentration of such ions. This is inconvenient for a number of reasons. First, the notation requires that we work with a pair of values (the factor and the exponent). In addition, the notation makes it difficult to quickly assess the level of acidity or basicity of a solution. And it becomes quite tiring to have to repeatedly say such words as *the concentration of* H^+ *is four point*

---

[12]The use of related values of q_1 and q_2 to work out the value of the constant, k, is common, not just in working out the value of the constant, k, in inverse proportion problems, but also in the determination of the values of constants in working with direct superposition, inverse superposition, direct proportion, and indeed any problem whose formulation includes a single constant.

[13]The switch from k to K_w is required if our equation is to conform to the conventions on communicating scientific formulas as k is an abstract constant (its particular value depends on the particular problem being studied) while K_w is a concrete constant whose value relates specifically to the dissociation of water. The difference between the use of k and K_w is similar to the difference between the use of x and, say, m to refer to an unknown: In abstract settings one may use x but in a concrete setting where one wishes to find the mass of an object, one should switch to the use of m.

We also refrain from the use of the numerical value 10^{-14} in the formula as it is a measured value and is, therefore, not exact. As explained earlier, the use of a symbol such as K_w implies that one is discussing the *exact* value of the dissociation constant for water and not some approximation to it.

five times ten to the power of negative six moles per litre! To deal with this issue, we re-write our concentrations as 1×10^{-e}, i.e., we set the factor to equal to 1, and adjust the exponent of 10 to get the desired concentration. As an example, if the concentration is 3.2×10^{-4}, we re-write this as $1 \times 10^{3.494\cdots}$ or, approximately, $1 \times 10^{3.5}$.[14]

This switch to the format 1×10^{-e} allows us to work with the concentrations of H^+ and OH^- ions by changing a single value: the exponent of 10. We also drop $1\times$ as is customary and simply write 10^{-e} to refer to such concentrations. We could, indeed, go further and retain only the positive part of the exponent, i.e., e. As an example, we could say that this solution has an e value of 3.5 for H^+ and this would imply that the concentration of H^+ is $1 \times 10^{-3.5}$ mol/L. As the reader can tell, the value of e is key to understanding the degree of acidity or basicity of a solution. For this reason, we provide it with a quantity name and a corresponding quantity symbol. The quantity name that is given to e is *power of hydrogen* as it is the exponent (power) that relates to the concentration of H^+ ions in the solution and the quantity symbol chosen for this quantity is pH with the lowercase letter p used as the symbol for *power* and the capital letter H used as the chemical symbol for hydrogen.[15]

With this convention on the naming of e and its associated quantity symbol, we can now write 10^{-pH}. Based on this agreement, we can write

$$[H^+] = 10^{-pH}$$

As an example, a solution with a pH value of 3.5 will have an H^+ concentration of

$$[H^+] = 10^{-3.5}$$
$$[H^+] = 3.16 \times 10^{-4} \text{ mol/L}$$

or about $0.000\,316$ mol/L.[16]

A similar argument as above leads to the definition of the quantity *power of hydroxide* or OH and the corresponding quantity symbol, pOH. Changing

---

[14]We will explain shortly how such exponents of 10 can be computed. For now, the reader can use a calculator to convert 3.2×10^{-4} and $1 \times 10^{3.494\cdots}$ to decimals to convince herself or himself that the two values are indeed the same.

[15]It is generally not a good idea to use a collection of symbols to refer to the value of a single quantity. Such constructs can become confusing as they leave the impression that the multiple symbols are being multiplied by one another. Here we follow the conventions in chemistry and adopt notations that are commonly used in this field.

[16]Note how we have simplified the manner in which we communicate the concentration of H^+ and OH^- ions: In place of the longer sentence *the concentration of H^+ ions in the solution is equal to three point one six times ten to the power of negative four moles per litre* we now say *the solution has a pH of* 3.5.

the significand in the value of [OH$^-$] to 1 and labelling the positive part of the exponent of 10 as pOH leads to the following formula

$$[\text{OH}^-] = 10^{-\text{pOH}}$$

Since typical numerical values of concentrations of H$^+$ and OH$^-$ ions fall between 1×10^{-14} to 1×10^{-1}, both pH and pOH range from 1 to 14.

It is important to note that a decrease in the value of pH by 1 increases [H$^+$] by a factor of 10. As an example, if the pH of a solution changes from 3 to 2, then [H$^+$] changes from 10^{-3} mol/L, or 0.001 mol/L in decimal notation, to 10^{-2} mol/L, or 0.01 mol/L in decimal notation. Since 0.01 is 10 times larger than 0.001, [H$^+$] has increased 10 times[17] which in turn implies a more acidic solution.

This argument can be extended to show that if pH decreases by 2, then [H$^+$] increases by a factor of 100, and if pH decreases by 3, then [H$^+$] increases by a factor of 1000 and so on. Indeed, moving from a pH of 7 in neutral water to a pH of 2 in gastric or stomach acid implies an increase in the number of H$^+$ ions by a factor of 100 000. This is quite substantial.[18]

We will now establish a relationship between pH and pOH.

As discussed above, a decrease in pH by 1 implies an increase in [H$^+$] by a factor of 10. Since the relationship between [H$^+$] and [OH$^-$] is inverse proportion, an increase in [H$^+$] by a factor of 10 implies a decrease in [OH$^-$] by a factor of $\frac{1}{10}$. This in turn implies an increase in pOH by 1.

Therefore, when pH decreases by 1, pOH increases by 1. A similar argument shows that any increase in pH by a certain amount is matched by a decrease in pOH by the same amount and any decrease in pH by a certain amount is matched by an increase in pOH by the same amount. The relationship between pH and pOH, then, is one of inverse superposition. The form of the equation relating pH and pOH, then, must match that of inverse superposition, i.e., either the standard form, $q_2 = k - q_1$ or the conservation form, i.e., $q_1 + q_2 = k$. In the context of acidic and basic solutions, the conservation form is often used, i.e., we write

$$\text{pH} + \text{pOH} = k$$

---

[17]Note that 10^{-3} (0.001 in decimal notation), is less than 10^{-2} (0.01 in decimal notation) so that a smaller pH value implies a larger concentration of H$^+$ ions. Negative exponents imply repeated division by the base so that 10^{-3} becomes $\frac{1}{1000}$ whereas 10^{-2} becomes $\frac{1}{100}$. For negative exponents, the larger the size of the exponent, the larger the number of times we divide by the base and the smaller the result.

[18]To get a feel for the size of increase, a change in pH from 7 to 2 implies that the ratio of water molecules to H$^+$ ions has changed from about 550 million to 1, to about 5500 to 1, i.e., the ratio of 550 million molecules of water to 1 H$^+$ ion has changed to the ratio of 5500 molecules of water to 1 H$^+$ ion.

where k is the constant of superpositionality. To find the value of k, we use a pair of related values for $[H^+]$ and $[OH^-]$. The simplest such pair is the one that relate to the concentration of these ions in pure water, i.e., pH=7 and pOH=7. This yields

$$pH + pOH = k$$
$$k = pH + pOH$$
$$k = 7 + 7$$
$$k = 14$$

with unit 1 as k is the result of the sum of two exponents.[19] The formula can be written as[20]

$$pH + pOH = -\log K_w$$

The measured value of K_w is 10^{-14} so that $-\log K_w$ becomes $-\log 10^{-14}$ which simplifies to $-(-14)$ or 14.

In chemistry textbooks this formula is often written as pH+pOH=14.[21]

The pH scale is a scale with values running from 1 to 14. When pH is 7, pOH is 7 as well so that there are an equal number of H^+ and OH^- ions in the solution. If pH decreases by 1 to 6, the concentration of H^+ ions increases 10 times, the concentration of OH^- ions decreases 10 times, pOH increases by 1 and the solution becomes acidic. The lower the pH, the higher the concentration of H^+ ions, the lower the concentration of OH^- ions and the more acidic the solution. Similarly, if pH increases by 1 to 8, the concentration of H^+ ions decreases 10 times, the concentration of OH^- ions increases 10 times, pOH decreases by 1 and the solution becomes basic. The higher the pH, the lower the concentration of H^+, the higher the concentration of OH^- and the more basic the solution.

A pH scale with values ranging from 1 to 14 along with a few examples of substances whose acidity or basicity falls in this range is given below.[22]

A pOH scale can be defined in a manner similar to a pH scale. Such a scale can be superimposed on a pH scale as shown below.

We let the reader work out the logic behind the pOH scale.

---

[19] An exponent must have unit 1 as it indicates the number of times the base should be multiplied or divided by itself.

[20] Here is a derivation:

$$[H^+][OH^-] = K_w$$
$$10^{-pH} \times 10^{-pOH} = K_w$$
$$10^{-pH-pOH} = K_w$$
$$-pH - pOH = \log K_w$$
$$pH + pOH = -\log K_w$$

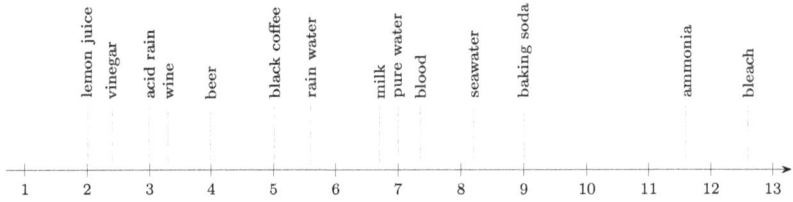

Figure 11.1.1: The pH scale

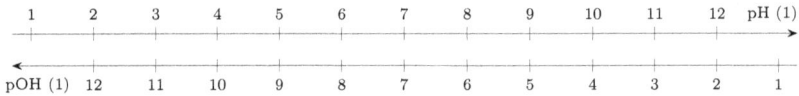

Figure 11.1.2: The pH and pOH scales

Let us summarize the four basic equations that are used in working with acidic and basic solutions.

$[H^+]$ concentration of H^+ ions in the solution (mol/L)
$[OH^-]$ concentration of OH^- ions in the solution (mol/L)
pH   the solution's power of hydrogen (1)
pOH   the solution's power of hydroxide (1)
K_w dissociation constant for water (mol^2/L^2)

$$[H^+][OH^-] = K_w$$

$$[H^+] = 10^{-pH}$$

$$[OH^-] = 10^{-pOH}$$

$$pH + pOH = -\log K_w$$

The first equation is based on theoretical explanations relating the values of $[H^+]$ and $[OH^-]$ and is backed by experimental data.

The second and third equations are based on definitions (i.e., agreements) on rewriting the concentrations of H^+ and OH^- as 10^{-e} and referring to the

---

[21] This is unfortunate as the use of a numerical value in a formula implies that the numerical value is exact. The notation $-\log K_w$ refers to the exact value of the quantity. In applications, this exact value may be replaced by its approximate measured value of 14.

[22] Other then the value given for pure water, the others are average values. As an example, the pH of acid rain can range from above 1 to below 5. The same can be said about the pH values of the other substances although the ranges are often smaller than the one attributed to acid rain.

positive part of the exponent of 10 as pH or pOH depending on whether we are talking about the concentration of H^+ or OH^- ions.

The last equation was derived mathematically using the first three equations.

As we showed earlier, the relationship between $[H^+]$ and $[OH^-]$ is inverse proportion which is a nonlinear relationship. The second and third equations are also nonlinear[23] whereas the last equation is linear as it expresses an inverse superposition relationship between the quantity symbols pH and pOH.

Before we end this section, we will present a few applications along with their solutions.[24]

Problem

A solution has an H^+ concentration of 2.8×10^{-6} mol/L. Calculate the concentration of OH^- ions in that solution.

Solution

$[H^+]$ concentration of H^+ ions in the solution (mol/L)
$[OH^-]$ concentration of OH^- ions in the solution (mol/L)
K_w dissociation constant for water $(\text{mol}^2/\text{L}^2)$

$$[H^+][OH^-] = K_w$$

$$[OH^-] = \frac{K_w}{[H^+]}$$

$$[OH^-] = \frac{1 \times 10^{-14}}{2.8 \times 10^{-6}}$$

$$[OH^-] = 3.571 \ldots \times 10^{-9} \text{ mol/L}$$

The concentration of OH^- is 3.57×10^{-9} mol/L.

We let the reader justify the solution above.

Problem

A solution has a pH value of 4.7. Calculate the concentration of H^+ ions in this solution.

---

[23]They are, in fact, inverse exponential relationships.

[24]For a discussion on the solution techniques used to solve nonlinear problems that involve exponents please see the companion textbook *Style in Technical Math* by the author.

Solution

$[H^+]$ concentration of H^+ ions in the solution (mol/L)
pH   the solution's power of hydrogen (1)

$[H^+] = 10^{-pH}$

$[H^+] = 10^{-4.7}$

$[H^+] = 1.995 \ldots \times 10^{-5}$ mol/L

The concentration of H^+ is 2.00×10^{-5} mol/L.

We let the reader justify the solution above.

Problem

A solution has an OH^- concentration of 5.8×10^{-4} mol/L. Calculate the solution's pOH.

Solution

$[OH^-]$ concentration of OH^- ions in the solution (mol/L)
pOH   the solution's power of hydroxide (1)

$[OH^-] = 10^{-pOH}$

$10^{-pOH} = [OH^-]$

$-pOH = \log[OH^-]$

$pOH = -\log[OH^-]$

$pOH = -\log(5.8 \times 10^{-4})$

$pOH = -(-3.236\ldots)$

$pOH = 3.236\ldots$

The solution has a pOH value of 3.2.

We let the reader justify the solution above.

Problem

A solution has a pH value of 2.9. Calculate the solution's pOH.

Solution

pH   the solution's power of hydrogen (1)
pOH   the solution's power of hydroxide (1)

$$pH + pOH = -\log K_w$$
$$pOH = -\log K_w - pH$$
$$pOH = -\log 10^{-14} - 2.9$$
$$pOH = -(-14) - 2.9$$
$$pOH = 14 - 2.9$$
$$pOH = 11.1$$

The solution has a pOH value of 11.1.

We let the reader justify the solution above.

Exercise Set 11.1

1. a. What happens to the concentration of OH^- ions in a solution if the concentration of H^+ ions increases by a factor of 10?
   b. What happens to the concentration of OH^- ions in a solution if the concentration of H^+ ions increases by a factor of 1000?
   c. What happens to the concentration of OH^- ions in a solution if the concentration of H^+ ions decreases by a factor of $\frac{1}{100}$?
   d. What happens to the concentration of H^+ ions in a solution if the concentration of OH^- ions increases by a factor of 1000?
   e. What happens to the concentration of H^+ ions in a solution if the concentration of OH^- ions increases by a factor of 4?

2. a. What happens to the concentration of H^+ ions in a solution if the pH value of the solution increases by 1?
   b. What happens to the concentration of H^+ ions in a solution if the pH value of the solution decreases by 2?
   c. What happens to the concentration of OH^- ions in a solution if the pOH value of the solution increases by 3?
   d. What happens to the concentration of OH^- ions in a solution if the pOH value of the solution decreases by 4?

3. a. What happens to the pOH value of a solution if its pH value increases by 1?
   b. What happens to the pOH value of a solution if its pH value decreases by 3?
   c. What happens to the pH value of a solution if its pOH value increases by 2?

    d. What happens to the pH value of a solution if its pOH value decreases by 3?

4. In each case determine whether the solution is acidic, basic, or neutral.

    a. The solution has a pH value of 8.7.
    b. The solution has a pH value of 2.1.
    c. The solution has a pH value of 7.0.
    d. The solution has a pOH value of 5.5.
    e. The solution has a pOH value of 7.0.
    f. The solution has a pOH value of 12.0.

5. a. The concentration of H^+ ions in a solution is 4.5×10^{-4} mol/L. Calculate the concentration of OH^- ions in that solution. The dissociation constant for water is 1×10^{-14} mol^2/L^2.

    b. The concentration of OH^- ions in a solution is 7.2×10^{-8} mol/L. Calculate the concentration of H^+ ions in that solution. The dissociation constant for water is 1×10^{-14} mol^2/L^2.

6. a. A solution has a pH value of 6.7. Calculate the solution's concentration of H^+ ions in that solution.

    b. A solution has a pOH value of 3.8. Calculate the solution's concentration of OH^- ions in that solution.

7. a. The concentration of H^+ ions in a solution is 6.1×10^{-9} mol/L. Calculate the solution's pH value.

    b. The concentration of OH^- ions in a solution is 1.4×10^{-12} mol/L. Calculate the solution's pOH value.

8. a. A solution has a pH value of 3.4. Calculate the solution's pOH. The dissociation constant for water is 1×10^{-14} mol^2/L^2.

    b. A solution has a pOH value of 9.7. Calculate the solution's pH. The dissociation constant for water is 1×10^{-14} mol^2/L^2.

11.2   Radiation and Biological Damage

In this section we will discuss matters of interest in relation to radiation problems and biological damage sustained as a result of exposure to radiation.[25]

Behaviour of Radioisotopes

Consider ^{131}I, a radioisotope of iodine used in the diagnosis and treatment of certain types of medical problems with the goiter gland. ^{131}I is a beta radioisotope with nuclear equation

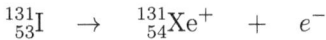

$$^{131}_{53}\text{I} \quad \rightarrow \quad ^{131}_{54}\text{Xe}^+ \quad + \quad e^-$$

Suppose we begin with 100 g of ^{131}I. As time goes on, ^{131}I gradually radiates and in doing so, decays into ^{131}Xe. If we were going to wait for about 8 days, we would find that half of the sample that we started with has decayed into ^{131}Xe, leaving us with only 50 g of ^{131}I.

It might seem logical to think that if another 8 days go by, the rest of ^{131}I will decay into ^{131}Xe and we will be left with no ^{131}I. This is not so. It may seem surprising but, if we were to wait another 8 days, we would find that we have lost half of the 50 grams that we were left with, i.e., we have lost only 25 g of ^{131}I, leaving us with 25 g of the radioisotope. And waiting for another round of 8 days would result in the loss of half of what we were left with at the end of the 16 days, leaving us with 12.5 g of the radioisotope.

We refer to 8 days as the **half-life** of ^{131}I. We use the quantity symbol t_h to refer to the value of the half-life of a radioisotope. Note that half-life is a time quantity[26] and, therefore, its unit must be a unit of time.

The behaviour described above is common to all radioisotopes although the actual value of the half-life is dependent on the particular type of the radioisotope. The range of half-lives is quite wide, with orders of magnitude ranging from 10^{-24} s and below to 10^{24} s and above.

Regardless of the actual duration of a half-life, it is the common behaviour of all radioisotopes that, after the passage of one half-life, half the sample that one starts with decays.

This behaviour can be turned into a mathematical model with relative ease. Consider, as an example, the case of starting with a certain mass of

---

[25]The concept of radiation and the manner in which radiation is generated is discussed at length in Appendix B and we refer the reader to this appendix for background information on radiation. The discussion that follows assumes this background on the part of the reader.

[26]Hence the choice of the symbol t to represent its value with.

a radioisotope. We use the quantity symbol m_0 to refer to the value of this initial mass. After 1 half-life (be it 0.002 s or 5000 years), we will be left with half of the initial mass, i.e., $\frac{1}{2} \times m_0$. After 2 half-lives we will be left with half of the mass that we had at the end of the first half-life, i.e., $\frac{1}{2} \times \frac{1}{2} \times m_0$. After 3 half-lives we will be left with half of the mass that we had at the end of the second half-life, i.e., $\frac{1}{2} \times \frac{1}{2} \times \frac{1}{2} \times m_0$, and so on. Note that the expression for the mass left after 1 half-life involves the single factor $\frac{1}{2}$ while the expression for mass left after 2 half-lives involves two such factors and the expression for mass left after 3 half-lives involves three such factors. We can, therefore, conclude that the mass of a radioisotope left after n half-lives[27] equals to

$$ m = \underbrace{\frac{1}{2} \times \frac{1}{2} \times \ldots \times \frac{1}{2}}_{n \text{ factors}} \times m_0 $$

where m represents mass left after n half-lives. Multiplying the numerators and denominators of these fractions leads to the following simplified formula.[28]

$$
\begin{array}{ll}
m_0 & \text{initial mass of radioisotope (kg)} \\
m & \text{final mass of radioisotope (kg)} \\
n & \text{number of half-lives passed (1)}
\end{array}
$$

$$ m = \frac{1}{2^n} m_0 $$

Problem

We start with 4.5 g of a radioisotope. What mass of the radioisotope will be left after 3.5 half-lives have gone by?

---

[27] The reader should differentiate between half-life and the number of half-lives. The former is a time quantity with a unit of time, e.g., 8 days, while the latter refers to the number of half-lives that have gone by. As an example, if the half-life of a radioisotope is 8 days, then, after 16 days we can say that 2 half-lives have gone by and after 24 days we can say that 3 half-lives have gone by and so on. Note that, for this radioisotope, the half-life remains fixed at 8 days while the number of half-lives that have passed may change from 2 to 3 to 4, etc.

[28] Any unit of mass may be used in this formula so long as the same unit of mass is used for both m_0 and m.

Solution

m_0 initial mass of radioisotope (g)
m final mass of radioisotope (g)
n number of half-lives passed (1)

$$m = \frac{1}{2^n} m_0$$

$$m = \frac{1}{2^{3.5}} \times 4.5$$

$$m = \frac{1}{11.313\ldots} \times 4.5$$

$$m = 0.397\ldots \text{ g}$$

There will be 0.40 g of the radioisotope left.

Here is an example with initial mass as unknown.

Problem

After 5.2 half-lives, 1.8 g of the radioisotope is left. What mass of the radioisotope did we start with?

Solution

m_0 initial mass of radioisotope (g)
m final mass of radioisotope (g)
n number of half-lives passed (1)

$$m = \frac{1}{2^n} m_0$$

$$\frac{1}{2^n} m_0 = m$$

$$m_0 = 2^n m$$

$$m_0 = 2^{5.2} \times 1.8$$

$$m_0 = 36.758\ldots \times 1.8$$

$$m_0 = 66.165\ldots \text{ g}$$

We started with 66 g of the radioisotope.

Here is a problem with the number of half-lives passed as unknown.

Problem

How many half-lives will have to pass until 78 g of a radioactive material deays to 5.9 g?

Solution

m_0 initial mass of radioisotope (g)
m final mass of radioisotope (g)
n number of half-lives passed (1)

$$m = \frac{1}{2^n} m_0$$

$$2^n = \frac{m_0}{m}$$

$$n = \log_2 \frac{m_0}{m}$$

$$n = \log_2 \frac{78}{5.9}$$

$$n = \log_2 13.220 \ldots$$

$$n = 3.724 \ldots$$

3.7 half-lives will have to pass for the mass of the radioisotope to drop to the indicated value.

If we know the half-life of a radioisotope and the time that has passed since we measured its mass, then we can compute the number of half-lives that have passed by dividing the latter by the former. As an example, if the half-life of a radioisotope is 8 days, then, after 16 days we can say that $\frac{16}{8}$ or 2 half-lives have gone by and after 24 days we can say that $\frac{24}{8}$ or 3 half-lives have gone by and so on. This leads to the following formula:[29]

t time passed (s)
t_h half-life of the radioisotope (s)
n number of half-lives passed (1)

$$n = \frac{t}{t_h}$$

which in turn leads to the second version of the formula for the computation

---

[29]Any unit of time may be used in this formula so long as the same unit of time is used for both t and t_h. In practice, it is common to use the unit of time that is used in the expression of the value of the half-life of the particular radioisotope under discussion. As an example, if the radioisotope under discussion has a half-life of 4 weeks, then weeks is used as the unit of time in the formula above.

of mass left after n half-lives:[30]

$$
\begin{array}{ll}
m_0 & \text{initial mass of radioisotope (kg)} \\
m & \text{final mass of radioisotope (kg)} \\
t & \text{time passed (s)} \\
t_h & \text{half-life of radioisotope (s)}
\end{array}
$$

$$
m = \frac{1}{2^{\frac{t}{t_h}}} m_0
$$

Problem

[131]I has a half-life of 8.05 days. How long will it take for 5.7 g of ^{131}I to decay to 4.2 g?

Solution

$$
\begin{array}{ll}
m_0 & \text{initial mass of radioisotope (g)} \\
m & \text{final mass of radioisotope (g)} \\
t & \text{time passed (days)} \\
t_h & \text{half-life of radioisotope (days)}
\end{array}
$$

$$
m = \frac{1}{2^{\frac{t}{t_h}}} m_0
$$

$$
2^{\frac{t}{t_h}} = \frac{m_0}{m}
$$

$$
\frac{t}{t_h} = \log_2 \frac{m_0}{m}
$$

$$
t = t_h \log_2 \frac{m_0}{m}
$$

$$
t = 8.05 \times \log_2 \frac{5.7}{4.2}
$$

$$
t = 8.05 \times \log_2 1.357 \ldots
$$

$$
t = 8.05 \times 0.440 \ldots
$$

$$
t = 3.546 \ldots \text{ days}
$$

It will take 3.5 days.

---

[30] Any unit of mass and any unit of time may be used in this formula so long as the same unit of mass is used for both m_0 and m and the same unit of time is used for both t and t_h.

Consider now the following problem.

Problem

What percentage of the mass of a radioisotope is left after the passage of 2.5 half-lives?

We now explain how such a problem can me modelled.

To begin, consider the case where we are asked to find the percentage of the radioisotope that is left after 1 half-life. As explained earlier, after 1 half life we are left with $\frac{1}{2}$ of the mass that we started with, i.e., $\frac{1}{2} \times m_0$. The answer to this problem, then, requires that we convert $\frac{1}{2}$ to a percentage, i.e., 50%. We can, therefore, say that, after the passage of 1 half-life, we are left with 50% of the radioisotope.

Consider now the case where we are asked to calculate the percentage of the radioisotope that is left after 2 half-lives. As explained earlier, after 2 half-lives we are left with $\frac{1}{2}$ of the mass that we had at the end of the first half-life, i.e., $\frac{1}{2} \times \frac{1}{2} \times m_0$. This simplifies to $\frac{1}{2^2} \times m_0$ or $\frac{1}{4} \times m_0$. The last expression reads as $\frac{1}{4}$ of m_0 so that, to answer the question, we can convert $\frac{1}{4}$ to percentage and report the value that we find, i.e., 25%.

A similar argument shows that after 3 half-lives we will be left with $\frac{1}{2} \times \frac{1}{2} \times \frac{1}{2} \times m_0$. This works out to $\frac{1}{2^3} \times m_0$ or $\frac{1}{8} \times m_0$. Since this reads as $\frac{1}{8}$ of m_0, we can compute the percentage of the mass that is left by converting $\frac{1}{8}$ to percentage and report what we compute.

Such problems, then, require that we work out the subexpression $\frac{1}{2^n}$ and convert its value to percentage. We have the following formula.[31,32]

---

[31]The justification given above for this formula is based on semantics. Here is a syntactic proof:

$$m = \frac{1}{2^n} m_0$$

$$\frac{m}{m_0} = \frac{1}{2^n}$$

$$r = \frac{1}{2^n}$$

Note that the fraction of mass left, i.e., r, may be computed by taking the ratio of mass left to initial mass, i.e., $\frac{m}{m_0}$.

[32]It is interesting to note that a chain such as $\frac{1}{2} \times \frac{1}{2} \times \frac{1}{2} \times$ and its simplified form $\frac{1}{2^3} \times$ or $\frac{1}{8} \times$ both have semantic significance. The former reads as

$$\frac{1}{2} \text{ of } \frac{1}{2} \text{ of } \frac{1}{2} \text{ of}$$

and is useful in making sense of successive losses in mass that are experienced by a radioisotope as half-lives go by. The latter reads as

r fraction of mass left (1)
n number of half-lives passed (1)

$$r = \frac{1}{2^n}$$

Here is a solution to the problem above.

r percentage of the mass of radioisotope left (1)
n number of half-lives passed (1)

$$r = \frac{1}{2^n}$$

$$r = \frac{1}{2^{2.5}}$$

$$r = \frac{1}{5.656\ldots}$$

$$r = 0.176\ldots$$

17.6% of the radioisotope is left after the passage of the specified number of half-lives.

Another version of the formula above uses the expression $\frac{t}{t_h}$ in place of n:[33]

r percentage of the mass of radioisotope left (1)
t time passed (s)
t_h half-life of radioisotope (s)

$$r = \frac{1}{2^{\frac{t}{t_h}}}$$

Biological Damage

A number of steps need to be taken for us to determine the biological damage that is sustained by living tissue that is exposed to radiation. First, we need to define a quantity and an associated unit that helps us measure the level of radiation that is emitted by the source of radiation. Next we need to define a quantity and an associated unit to measure how much of this radiation is absorbed by living tissue. And third we need to define a quantity and an associated unit to measure the damage that is sustained by the tissue.

$\frac{1}{8}$ of

and is useful when one wishes to compute the fraction of mass that is left.

[33] Any unit of time may be used in this formula so long as the same unit of time is used for both t and t_h.

To measure the amount of radiation emitted by a radioactive substance, we can measure the number of disintegrations that it experiences in a unit of time. This quantity is called **activity** with quantity symbol A. The SI unit of activity is the *becquerel* with unit symbol Bq. A radioactive substance with an activity of 1 Bq experiences 1 disintegration per second. We have the following definition.

A activity of the radioactive substance (Bq)
N number of disintegrations (1)
t time (s)

$$A = \frac{N}{t}$$

To measure the amount of radiation that is absorbed by exposed tissue we can use the energy that the radiation deposits in the tissue per unit mass of the tissue.[34] This quantity is called **absorbed dose** with corresponding quantity symbol D. Based on the definition of absorbed dose given above, the SI unit of absorbed dose is J/kg, a unit that is called *gray* with corresponding unit symbol, Gy. 1 Gy is defined as the generation of 1 J of heat by radiation in 1 kg of the target tissue. We have the following.

D absorbed dose (Gy)
E heat generated by absorbed radiation (J)
m mass of target tissue (kg)

$$D = \frac{E}{m}$$

Absorbed dose plays a role in the calculation of biological damage but experimental data show that it is not the only player as damage sustained depends also on the type of radiation. As an example, it is known that an absorbed dose of 1 Gy from an alpha radioisotope causes 20 times more damage to tissue than an absorbed dose of 1 Gy from a beta radioisotope. To measure biological damage, then, one must incorporate a weight that incorporates the effect of the type of radiation that the target tissue is exposed to. This weight is called the **radiation weighting factor** with corresponding

---

[34]Since it is difficult to measure the number of hits that the tissue receives directly, it is common to measure the amount of energy that is deposited in the target tissue as this energy is equal to the sum total of all the energies that are deposited in the target tissue by the individual hits and, therefore, relates to the number of hits. This energy appears as heat so that it can be measured easily. We can, therefore, measure the absorbed dose as the amount of heat energy that is deposited in the target tissue. However, it is also important to note the mass of the target tissue. To explain why, consider two tissue samples, one small and one large. While absorption of the same amount of radiation by the two tissue samples causes the same amount of damage overall, the effect is more pronounced on the smaller tissue as the damage is concentrated in a smaller area.

quantity symbol, w_R. For beta and gamma radiations, the radiation weighting factor is set equal to 1. For alpha radiation, it is set equal to 20.

The quantity measuring the extent of damage that is sustained is called **equivalent dose** with quantity symbol H. The SI unit of equivalent dose is the *sievert* with unit symbol Sv. 1 Sv is the damage sustained as a result of the absoroption of 1 J of energy deposited per kilogram of tissue but recall that this value is weighted to include the effect of the type of the radiation.[35] The formula relating the equivalent dose to absorbed dose is

H equivalent dose (Sv)
D absorbed dose (Gy)
w_R radiation weighting factor (1)

$$H = w_R D$$

Yet another quantity, called the **effective dose** with quantity symbol H_T, incorporates the effect of tissue type in assessing the damage through the incorporation of a weight, called the **tissue weighting factor** with symbol, w_T, that is applied to the effective dose.[36] The formula is[37]

H_T effective dose (Sv)
H equivalent dose (Sv)
w_T tissue weighting factor (1)

$$H_T = w_T H$$

Examples of the values of w_T are 0.12 for bone marrow, colon, lungs, and stomach, 0.05 for bladder, breasts, liver, esophagus, and thyroid, and 0.01 for skin.

According to data published by the Canadian Nuclear Safety Commission, the average annual effective dose due to background radiation in Canada is about 1.8 mSv. In comparison, data published by the US Food and Drug Administration notes that a typical chest X-ray results in an effective dose of about 0.1 mSv while a typical CT scan of the chest results in an effective dose of about 7 mSv. Astronauts working on the International Space Station are exposed to an annual effective dose of about 320 mSv. An effective dose of about 500 mSv or 0.5 Sv received over a short period of time can result in symptoms of radiation sickness. An effective dose of about 2 Sv received over a short period of time causes severe radiation poisoning and can be fatal.

[35]This means that an equivalent dose of 1 Sv causes the same amount of damage irrespective of the type of the radiation.

[36]This means that an effective dose of 1 Sv causes the same amount of damage irrespective of the type of tissue that it targets.

[37]Note that the unit of effective dose is set to be the same as the unit of equivalent dose. This is unfortunate as it makes it impossible to tell whether the unit refers to equivalent or effective dose. A much better course of action would have been to choose a new unit for effective dose in much the same way that sievert was invented to differentiate equivalent dose from absorbed dose.

Exercise Set 11.2

1. We have 8.7 g of a radioisotope. What mass of the radioisotope will be left after 3.5 half-lives?

2. What mass of the radioisotope will be left after 0.5 half-lives if we start with 6.2 g of the radioisotope?

3. We started with 20.6 g of a radioisotope and now we have 12.7 g of the radioisotope left. How many half-lives have passed?

4. How many half-lives will have to pass until 138 g of a radioisotope decays to 4.5 g?

5. After 3.8 half-lives we have 0.8 g of the radioisotope left. What mass of the radioisotope did we start with?

6. After 0.75 half-lives we have 12.3 g of the radioisotope left. What mass of the radioisotope did we start with?

7. A radioisotope has a half-life of 5.7 days. Calculate the number of half-lives corresponding to the passage of 12.5 days.

8. The half-life of a radioisotope is 4.5 years. 0.5 years have passed since a sample of the radioisotope was generated. How many half-lives does this correspond to?

9. Calculate the duration of time that corresponds to the passage of 9.3 half-lives for a radioisotope with a half-life of 7.7 s.

10. 0.9 half-lives have passed since we generated a sample of a radioisotope. What duration of time does this correspond to? The radioisotope has a half-life of 12.8 min.

11. 4.5 half-lives of a radioisotope correspond to 13.7 min. What is the half-life of the radioisotope?

12. 9.2 days have passed since we generated a sample of a radioisotope. This corresponds to 0.8 half-lives. What is the half-life of the radioisotope?

13. We started with 7.8 g of ^{24}Na, a radioisotope of sodium. What mass of ^{24}Na will be left after 48 h? ^{24}Na has a half-life of 15 days.

14. We have 28.7 g of ^{32}P, a radioisotope of phosphorus with a half-life of 14.2 days. What mass of ^{32}P will be left after 2 days?

15. We placed 132 kg of ^{134}Cs in storage and now we have 24.54 g of the radioisotope left. How long has the sample been in storage? The half-life of ^{134}Cs is 2.06 years.

16. ^{241}Am has a half-life of 432 years. How long will it take for 0.5 g of ^{241}Am to decay to 0.49 g?

17. We started with 14.5 g of ^{7}Be and after 90 days we have 4.47 g of the radioisotope left. Calculate the half-life of ^{7}Be.

18. 14.8 g of ^{99m}Tc decays to 5.23 g in 9 h. What is the half-life of ^{99m}Tc?

19. After 0.25 years we are left with 12.8 g of ^{22}Na. What mass of ^{22}Na did we start with? ^{22}Na has a half-life of 2.6 years.

20. After 24 h we are left with 2.69 g of ^{66}Ga, a radioisotope of gallium with a half-life of 9.4 h. What mass of ^{66}Ga did we start with?

21. What percentage of the mass of a radioisotope will be left after 1.75 half-lives have passed?

22. Calculate the percentage of the mass of a radioisotope that will be left after 12.7 half-lives have passed.

23. How many half-lives does it take for a sample of a radioisotope to decay to 30% of its original mass?

24. How many half-lives will it take for a sample of a radioisotope to decay to 10% of its initial mass?

25. What percentage of the mass of a radioisotope is left after 8.5 months? The radioisotope has a half-life of 3.2 months.

26. What percentage of the mass of a radioisotope is left after 12.9 h if the radioisotope has a half-life of 15 h?

27. A radioisotope has a half-life of 9.1 days. How many days does it take for its mass to drop to 40% of its initial value?

28. How long will it take for the mass of a sample of a radioisotope to decay to 90% of its initial value? The radioisotope has a half-life of 132 s.

29. It takes 5.5 h for the mass of a sample of a radioisotope to reduce to 15% of its initial value. What is the half-life of the radioisotope?

30. After 3.8 days we are left with 21.8% of the mass of a radioisotope. Calculate the radioisotope's half-life.

31. A sample of radioactive material experiences 126 disintegrations in 14.5 s. What is the sample's activity?

32. A sample of a radioactive material has an activity of 63 Bq. How many disintegrations does such a sample experience in 38.5 s?

33. 0.11 J of energy were absorbed in the form of heat from radiation by living tissue having a mass of 0.38 kg. Calculate the tissue's absorbed dose.

34. Calculate the heat absorbed by 0.042 kg of living tissue if the tissue experiences an absorbed dose of 0.0175 Gy.

35. A sample of tissue experiences an absorbed dose of 0.43 Gy due to exposure to beta rays. Calculate the equivalent dose received by the tissue. The radiation weighting factor for beta rays is set to 1.

36. Calculate the radiation weighting factor for alpha rays if an absorbed dose of 0.000 75 Gy results in an equivalent dose of 0.015 Sv.

37. A sample of liver receives an equivalent dose of 0.46 Sv. Calculate the

effective dose received by the tissue. The tissue weighting factor for liver is set to 0.05.

38. A sample of bone marrow receives an effective dose of 0.002 83 Sv. Calculate the equivalent dose received by the tissue. The tissue weighting factor for bone marrow is 0.12.

Part VI

Formulas

In this part of the book we will discuss
the genesis of formulas and the manner in which
one can use visual cues to relate the values of
the quantities that appear in a formula.

Chapter 12
Genesis

Formulas are general statements about how the values of certain quantities are related to the values of other quantities.

Some formulas arise as a result of the manner in which we define new quantities by relating their values to the values of one or more existing quantities. Formulas that relate to theoretical science often fall into this category.

Other formulas arise as a result of observations into the manner in which the values of certain quantities are naturally related to the values of other quantities. Matters that relate to experimental science often fall into this category.

An example of a defined formula in physics is the formula

v speed of an object (m/s)
d distance covered (m)
t travel time (s)

$$v = \frac{d}{t}$$

We see this formula as the definition of the quantity *speed*. A *definition* introduces a new quantity that we find useful in some way and in doing so, relates its value to the values of quantities that we are familiar with. In the case of the definition above, the new quantity introduced is called *speed* and it is defined as the rate of change of distance with respect to time.[1]

An example of a formula that is based on the results of experimental data

---

[1] Awareness for the need to define a new quantity is based on experience: If one needs to refer to the result of a certain calculation frequently, then one might choose to define a new quantity and asociate the results of such calculations to its value. As an example, if we find that we need to repeatedly compute the rate at which distance is covered with respect to time by dividing the covered distance by travel time, then we might choose to refer to such results as *speed* rather than *the rate of coverage of distance with respect to time.*

is one that relates the values of the various quantities that influence the state of an ideal gas, i.e., its amount, pressure, volume and temperature. Such formulas are of course approximate as they rely on the results of experimental data which are themselves approximations. Theoretical explanations can, however, change these formulas into exact formulas. In the case of the example above, the theoretical formula is called the *Ideal Gas Equation of State* and takes the following form.

p pressure of the gas (Pa)
V volume of the container (m^3)
n amount of the gas (mol)
T temperature of the gas (K)
R universal gas constant $(\text{J}/(\text{mol} \cdot \text{K}))$

$$pV = nRT$$

The origins of this formula are in experimental data that attempted to relate changes in the value of one quantity in the formula to changes in the value of another quantity in the formula. As an example, experiments run by Robert Boyle and others showed that, if everything else (i.e., the temperature of the gas and its amount) is kept constant, then the pressure of a gas is inversely proportional to the volume of its container. This led to the conservation formulation $pV = k$. Other experiments done by Jacques Charles showed that, if everything else (i.e., the pressure and amount of the gas) is kept constant, then the volume of a container is directly proportional to the temperature of the gas that it contains. This led to the standard formulation $V = kT$. The equation $pV = nRT$ summarizes all such relationships with R acting as the constant.

The experimental relationships above were discovered before any explanations for them were given. Of the competing explanations, the one that is currently accepted is the one that relates the observed relationships between the quantities in the formula to the state of movement of the molecules of which the sample of gas is made. This theory, called the *Kinetic Theory of Gases*, claims that the atoms and molecules within an object are not stationary but that they move around and about at random speeds and in random directions. According to this theory, the phenomena (e.g., existence of pressure) and relationship between the quantities that these phenomena represent (e.g., inverse proportion relationship between pressure and volume) are all rooted in the random movement of the molecules within the gas.

As an example, it is this random movement of the molecules that generates pressure: The random movement implies that the molecules often collide with each other and the walls of the container in which they are placed. Pressure is the manifestation of the force that is exerted by the molecules that hit and bounce off of the walls of the container.

In addition, the theory can show that the temperature of a given gas is related to the average speed with which its molecules move. The higher the temperature, the faster the molecules within the gas move around and about.[2]

As an example of how the Kinetic Theory of Gases explains the inverse proportion relationship between pressure and volume, consider the case where we have 10 molecules in a container. We begin by assuming that the amount of the gas and its temperature are both constant.[3]

Suppose now that we reduce the volume of the container to half its original size. Since there are now twice as many molecules per unit volume of the container, there will be twice as many hits on a unit area of the walls of the container which results in twice the pressure. Therefore, half the volume implies twice the pressure. We can repeat this argument, this time reducing the volume of the container to one third of its initial size. The argument shows that we will now have three times as many hits on a unit area of the walls of the container which drive up the pressure to three times its original value. Repeated arguments of this type or their generalizations show that the relationship between the value of the pressure of the gas and that of the volume of its container is one of inverse proportion.

Theoretical explanations may lag or lead experimental data: One may first perform experiments to relate the values of two quantities (e.g., use experimental results that show pressure is inversely proportional to volume) and then try to come up with an explanation for the observed relationship, or alternatively, one may come up with an explanation as to why the relationship between two quantities must be of a certain sort and then perform experiments to see whether the theoretical explanation is valid. Since experimental results have the final say, it is often easier to follow the former route. The latter is used when technological limitations prevent us from performing experiments or when the cost of performing experiments is prohibitive.

A theoretical explanation that have not been thoroughly tested is referred to as a hypothesis. Once a hypothesis has been extensively tested and verified

---

[2]Note that this also provides an explanation for the existence of the coldest possible temperature. The slower the molecules move around and about, the lower the temperature. We reach the coldest temperature when molecules stop moving. This happens at -273.15 °C or 0 K.

[3]Note that it is important to keep the amount of gas and its temperature constant as changes in the values of these quantities can have an effect on the value of the pressure of the gas. This can be shown to be so experimentally but the Kinetic Theory of Gases can also explain these effects. As an example, if we increase the amount of gas, then there will be more molecules in the container resulting in more hits per unit area of the wall of the container which increases the pressure. And if we increase the temperature of the gas, then its molecules will move faster and, therefore, they hit the walls of the container harder and this increases the pressure. To relate the value of pressure to that of volume, we need to make sure that changes in pressure are due to changes in volume only, hence the need to keep the amount and temperature constant.

by such tests, the hypothesis is elevated to the status of a theory. Since scientific theories have been tested extensively and verified by such tests, we have a lot of confidence in their validity.

Note also that it is often possible to come up with multiple theoretical explanations that are able to relate the values of scientific quantities in ways that are supported by experimental data. In such cases the simplest explanation is chosen as the likely correct explanation.

Before ending this section we wish to also briefly address the question as to whether scientific formulas can ever be wrong. This is of course possible but, since the final say rests in the hands of experimental results, the only way this can happen is for the experimental setup to have flaws and that these flaws must have gone unnoticed by the tens of thousands of scientists who have set them up in one way or another for one reason or another. While it is always possible that there may be such systemic flaws, the chances of them occurring are exceedingly small.[4]

In the absence of such flaws, scientific theories are quite sound although they evolve. Since scientific theories are justified based on the results of experimental data, their validity is limited to the limits within which they have been tested. There are no guarantees that the relationships that are found within such limits continue to hold outside those limits as new players may enter the field as we move beyond such limits. In this sense, newer theories expand the capabilities of older ones by making them applicable to a wider range of values.

As an example, Newton's Laws have been used for over two hundred years in a wide range of applications that require a balance of forces. They continue to be used extensively today from the design of buildings and bridges to the balance of airplanes as they fly and the design of engines that propel them. However, the formulas were never tested under conditions when the speeds that are involved are close to the speed of light until much later. It would, therefore, have been dangerous for us to assume that Newton's Laws should hold for all possible values of the variables that go into their models and, as you can see, they don't. An extension of Newton's Laws was given by Einstein Theory of Relativity. The new formulas can be used to describe the relationships between the various quantities for a wider range of speeds. At low speeds, however, practitioners still follow Newton's Laws as the effects of the players that become significant at higher speeds are quite insignificant when speeds are low.

Newer theories do carry philosophical implications with them however. As an example, while we continue to use Newton's formulas in the design

---

[4]Note that such flaws, if they exist, would have a major effect on science as the theories that are devised around erroneous data would naturally be erroneous leading us in the wrong direction.

of our buildings and bridges, Einstein's Theory of Relativity which extends Newton's Laws has shown that the fabrics of space and time may not be as simple as we thought they were.

Chapter 13
Relating the Values of Quantities that Appear in a Formula

As we saw in the previous chapter, the relationship between the values of the various quantities that play a role in a problem can be expressed as a formula. However, formulas can also help us remember these relationships by providing visual cues that allow us to classify problem type.

As an example, consider the following formula

p price ($)
r rate of tax (1)
a amount of tax ($)

$$a = rp$$

This formula should be familiar to the reader by now. The formula relates the values of the quantities *amount of tax*, *rate of tax* and *price*. The formula may be used to compute the value of any of the three quantities *amount of tax*, *rate of tax* or *price*, given the values of the other two.

However, in addition to the above, we can tell by looking at the form of this formula that the amount of tax is directly proportional to the rate of tax as well as the price. We can also tell that the price is inversely proportional to the rate of tax.

How can we tell all of this by just looking at a formula?

In order to be able to relate the values of the quantities that appear in a formula, we use the following structures as guides:[1,2]

---

[1] The abstract counterparts of these equations using x and y in place of q_1 and q_2 are discussed at length in the companion textbook *Style in Technical Math* by the author. The reader should read this material before continuing further.

[2] The formulas listed above correspond to the basic applications of math that involve addition, subtraction, multiplication and division. There are others of course: $q_2 = q_1^k$, $q_2 = \sqrt[k]{q_1}$, $q_2 = k^{q_1}$, etc. We will not discuss these other forms in this textbook to any great extent.

Standard Forms of Equations for Direct and Inverse Superposition and Direct and Inverse Proportion

Direct superposition: $q_2 = k + q_1$

Inverse superposition: $q_2 = k - q_1$

Direct proportion: $q_2 = kq_1$

Inverse proposition: $q_2 = \dfrac{k}{q_1}$

Conservation Forms of Equations for Direct and Inverse Superposition and Direct and Inverse Proportion

Direct superposition: $q_2 - q_1 = k$

Inverse superposition: $q_2 + q_1 = k$

Direct proportion: $\dfrac{q_2}{q_1} = k$

Inverse proposition: $q_2 q_1 = k$

In all the equations above, k represents a **constant**, i.e., a quantity whose value is, for a given problem, fixed.

In addition to the above, we will use the following form to classify a relationship as linear:

Linear Form

Linear: $q_2 = mq_1 + b$

where m and b are constants with m representing the slope of the associated line and b representing the q_2-coordinate of the q_2-intercept.[3]

We now take a closer look at each category.

Direct Superposition

The relationship between two quantities is of type direct superposition if, an increase/decrease in the value of one of them by a certain amount forces a

---

[3]The choice of q_1 and q_2 in place of x and y reflects our insistence that, in the context of applied problems, these symbols represent the values of quantities.

corresponding increase/decrease in the value of the other by the same amount. Such relationships have the form of

$$q_2 = k + q_1, \qquad k \in \mathbb{R}$$

The side condition, $k \in \mathbb{R}$, formally states that k must be a known number, i.e., a member of $\mathbb{R}$.

As an example of how this equation forces a relationship of type direct superposition on q_1 and q_2, suppose q_1 increases by 3 units. Since q_1 is being added, we will be adding 3 more units to k, raising the total, i.e., q_2, by 3 units as well.

We can therefore, say that change in the value of one quantity is matched by change in the value of the other quantity, i.e.,[4]

$$\Delta q_2 = \Delta q_1$$

Direct superposition is symmetric. This means that, if q_2 is directly superpositional to q_1, then q_1 is directly superpositional to q_2. This can be seen through a rearrangement of the equation for direct superposition as shown below.

$$q_2 = k_1 + q_1 \qquad q_2 \text{ is directly superpositional to } q_1$$
$$k_1 + q_1 = q_2$$
$$q_1 = -k_1 + q_2$$
$$q_1 = k_2 + q_2 \qquad q_1 \text{ is directly superpositional to } q_2$$

Note that k_1 is assumed to be in $\mathbb{R}$ which implies that k_2 is also a member of $\mathbb{R}$ as k_2 is equal to $-k_1$.

Since, as noted above, in a relationship of type direct superposition an increase/decrease in the value of one quantity by a certain amount forces a corresponding increase/decrease in the value of the other quantity by the same amount, the difference between the values of the two quantities must remain constant. Mathematically we can show this through the rearrangement

$$q_2 = k + q_1$$
$$q_2 - q_1 = k$$

which leads from the standard form of the equation for direct superposition to the conservation form of the equation for direct superposition.[5]

As an example of a direct superposition problem, consider the formula that relates the value of the temperature expressed in degrees Celsius to its

---

[4]The claims made in this textbook are based on semantics. Syntactic justification for these claims are given in the companion textbook *Style in Technical Math* by the author.

[5]Both forms are used in practice as some problems naturally favour one while others favour the other.

value expressed in kelvins, i.e.,

t temperature (°C)
T temperature (K)
T_0 temperature at which water freezes under standard conditions (K)

$$T = T_0 + t$$

In this formula T_0 represents the value of temperature in kelvins at which water freezes under standard conditions. This value is defined to be 273.15 K and is a constant.

The formula tells us that, to calculate the value of temperature in kelvins, we can add the value of the temperature in degrees Celsius to the constant 273.15.

In addition, since the equation has the form of $q_2 = k+q_1$, the relationship between temperature in degrees Celsius and temperature in kelvins is one of direct superposition. This means the following:

Since in a relationship of type direct superposition an increase/decrease in the value of one quantity by a certain amount forces a corresponding increase/decrease in the value of the other quantity by the same amount, we can say that an increase/decrease in the value of temperature in degrees Celsius by a certain amount forces a corresponding increase/decrease in the value of the temperature in kelvins by the same amount. As an example, if the temperature in degrees Celsius decreases by 14.8 °, then we can conclude that the temperature in degrees kelvin falls by 14.8 K.

Furthermore, since direct superposition is symmetric, we can conclude that an increase/decrease in the value of the temperature in kelvins by some amount forces a corresponding increase/decrease in the value of the temperature in degrees Celsius by the same amount. As an example, if the temperature in kelvins decreases by 14.8 K, then we can conclude that the temperature in degrees Celsius falls by 14.8 °C.

A side implication of this relationship is that the size of the unit is the same on both scales. This is a useful observation.

Finally, since the conservation form of the equation for direct superposition states that the difference between the values of the two quantities is constant (i.e., $q_2 - q_1 = k$), we can conclude that the difference between the values of the two temperatures is always constant. This is indeed the case as shown in $T - t = T_0$ with the value of the constant, T_0, set at 273.15 K by definition.

The discussion above on the form of the model for direct superposition and its properties can be adapted to the models and properties of other forms.

In what follows, we will point to the main conclusions and ask the reader to provide justification for these conclusions.

Inverse Superposition

The relationship between two quantities is of type inverse superposition if an increase/decrease in the value of one of them by a certain amount forces a corresponding decrease/increase in the value of the other by the same amount. Such relationships have the form of

$$q_2 = k - q_1, \qquad k \in \mathbb{R}$$

The side condition, $k \in \mathbb{R}$, formally states that k must be a known number, i.e., a member of $\mathbb{R}$.

In a relationship of type inverse superposition, change in the value of one quantity is related to change in the value of the other quantity through the relation

$$\Delta q_2 = -\Delta q_1$$

Inverse superposition is symmetric.

Since, as noted above, in a relationship of type inverse superposition an increase/decrease in the value of one quantity by a certain amount forces a corresponding decrease/increase in the value of the other quantity by the same amount, the sum of the values of the two quantities must remain constant, i.e.,

$$q_2 + q_1 = k$$

which leads from the standard form of the equation for inverse superposition to the conservation form of the equation for inverse superposition.[6]

As an example of an inverse superposition problem, consider the problem of travelling from Toronto to Montreal, a distance of 542 km. If a distance d_c has already been covered, the remaining distance, d_r, may be computed through

d distance between Toronto and Montreal (km)
d_c distance already covered (km)
d_r remaining distance (km)

$$d_r = d - d_c$$

---

[6]Both forms are used in practice as some problems naturally favour one while others favour the other.

In this formula d represents the distance from Toronto to Montreal. This value is measured to be 542 km and is a constant.

The formula tells us that, to calculate the remaining distance, we can subtract the distance that we have covered from the total distance, a constant with a value of 542 km.

In addition, since the equation has the form of $q_2 = k - q_1$, the relationship between the remaining distance and distance covered is one of inverse superposition. This means the following:

Since in relationships of type inverse superposition an increase/decrease in the value of one of the quantities by a certain amount forces a corresponding decrease/increase in the value of the other quantity by the same amount, we can say that an increase/decrease in the value of the covered distance by a certain amount forces a corresponding decrease/increase in the value of the remaining distance by the same amount. As an example, if we cover an additional 50 km, then the remining distance decreases by 50 km.

Furthermore, since the relationship is symmetric, we can conclude that an increase/decrease in the value of the remining distance by a certain amount forces a corresponding decrease/increase in the value of distance covered by the same amount. As an example, if the remaining distance decreases by 50 km, then we must have covered an additional 50 km.

Finally, since the conservation form of the equation for inverse superposition states that the sum of the values of the two quantities is constant (i.e., $q_2 + q_1 = k$), we can conclude that the sum of distance covered and remining distance is constant. This is indeed the case as shown in $d_r + d_c = d$ with the value of the constant, d, representing the value of the distance from Toronto to Montreal, measured as 542 km.

Direct Proportion

The relationship between two quantities is of type direct proportion if scaling the value of one of them by a certain factor forces a corresponding scaling of the value of the other by the same factor. Such relationships have the form of

$$q_2 = kq_1, \quad k \in \mathbb{R}$$

The side condition, $k \in \mathbb{R}$, formally states that k must be a known number, i.e., a member of $\mathbb{R}$.

In a relationship of type direct proportion, change in the value of one quantity is related to change in the value of the other quantity through the relation

$$\Delta q_2 = k\Delta q_1$$

i.e., the change in q_2 is k times the change in q_1.

Direct proportion is symmetric.

Since, as noted above, in a relationship of type direct proportion scaling the value of one quantity by a certain factor forces a corresponding scaling of the value of the other quantity by the same factor, the quotient of the values of the two quantities must remain constant, i.e.,

$$\frac{q_2}{q_1} = k$$

which leads from the standard form of the equation for direct proportion to the conservation form of the equation for direct proportion.[7]

As an example of a direct proportion problem, consider the problem of travelling along a path at a constant speed of 120 km/h. For such a problem, distance covered is related to the speed and travel time through

d distance covered (km)
v speed (km/h)
t travel time (h)

$$d = vt$$

In this formula v represents the speed of the car which is set to the constant value, 120 km/h.

The formula tells us that, to calculate the distance that we have covered, we can multiply the travel time by the speed.

In addition, since the equation has the form of $q_2 = kq_1$, the relationship between distance covered and travel time is one of direct proportion. This means the following:

First, since in relationships of type direct proportion scaling the value of one of the quantities by a certain factor forces a corresponding scaling of the value of the other quantity by the same factor, we can say that scaling the value of travel time by a certain factor forces a corresponding scaling of distance covered by the same factor. As an example, if we drive twice as long, we cover twice as much ground.

Furthermore, since the relationship is symmetric, we can conclude that scaling the value of the distance that is covered by a certain factor, forces the

---

[7]Both forms are used in practice as some problems naturally favour one while others favour the other.

scaling of travel time by the same factor. As an example, if we covered three times as much ground, we must have travelled three times as long.

Finally, since the conservation form of the equation for direct proportion states that the quotient of the values of the two quantities is constant (i.e., $\frac{q_2}{q_1} = k$), we can conclude that the quotient of distance covered and travel time is constant (i.e., $\frac{d}{t} = v$). This is indeed the case with the constant representing the speed of 120 km/h.

Inverse Proportion

The relationship between two quantities is of type inverse proportion if scaling the value of one of the quantities by a certain factor forces a corresponding scaling of the value of the other quantity by the inverse of the same factor. Such relationships have the form of

$$q_2 = \frac{k}{q_1}, \qquad k \in \mathbb{R}$$

The side condition, $k \in \mathbb{R}$, formally states that k must be a known number, i.e., a member of $\mathbb{R}$.

Inverse proportion is symmetric.

Since, as noted above, in a relationship of type inverse proportion scaling the value of one of the quantities by a factor forces a corresponding scaling of the value of the other quantity by the inverse of the same factor, the product of the values of the two quantities must remain constant, i.e.,

$$q_2 q_1 = k$$

which leads from the standard form of the equation for inverse proportion to the conservation form of the equation for inverse proportion.[8]

As an example of an inverse proportion problem, consider the problem of calculating the power of a device. Power is defined at the rate of use of energy per unit time with the SI unit of J/s, a unit that is called a watt, with corresponding unit symbol, W. 1 W of power refers to the usage of 1 J of energy in 1 s. We have the following formula.

P device's power (W)
E energy used (J)
t duration of use (s)

$$P = \frac{E}{t}$$

---

[8]Both forms are used in practice as some problems naturally favour one while others favour the other.

Consider now the situation where an amount of energy of 1200 J has been used. This turns E in the formula above into a constant.

The formula tells us that, to calculate the power that a device uses, we can divide the amount of energy that is used by the duration of use of that energy.

In addition, since the equation has the form of $q_2 = \frac{k}{q_1}$, the relationship between power and duration of use is one of inverse proportion. This means the following:

First, since in relationships of type inverse proportion scaling the value of one of the quantities by a certain factor forces a corresponding scaling of the value of the other quantity by the inverse of the same factor, we can say that scaling the value of duration of use of energy by a factor forces a corresponding scaling of the power of the device by the inverse of the same factor. As an example, if it takes twice longer for one device to use the same amount of energy used by another device, then the first device has half the power of the second device. It does make sense that this should be so.

Furthermore, since the relationship is symmetric, we can conclude that scaling the value of the power by a certain factor, forces the scaling of the duration of use of energy by the inverse of the same factor. As an example, if one device has twice the power of another device, then the it uses the same amount of energy in half the time.

Finally, since the conservation form of the equation for inverse proportion states that the product of the values of the two quantities is constant (i.e., $q_2 q_1 = k$), we can conclude that the product of power and duration of use of energy is constant. This is indeed the case with the constant representing the amount of energy that is used, i.e., 1200 J in the problem above.

Linear Relationships

The relationship between two quantities is linear if their model has the form of

$$q_2 = mq_1 + b, \qquad m,b \in \mathbb{R}$$

The side condition, $m,b \in \mathbb{R}$, formally states that m and b must be known numbers, i.e., members of $\mathbb{R}$.

Linearity is symmetric.

Linear relationships may be thought of as those where the value of one quantity can be computed by adding an initial, fixed amount (b) to a variable amount (mq_1) with the variable amount being directly proportional to the value of the other quantity (mq_1 is directly proportional to q_1).

As an example of a linear relationship, consider the problem of financing a purchase whereby we pay an initial down payment of $750 followed by monthly payments at $125/month. We can compute the amount of money that has been repaid at any point in time using the following formula.

> a amount of money repaid ($)
> t time passed (months)
>
> $a = 125t + 750$

The formula tells us that, to calculate the amount of money that has been repaid, we can add the down payment ($750) to the total amount of monthly payments ($125t$), and that the latter can be calculated by multiplying the monthly payment of $125 by the time passed.

In addition, since the equation has the form of $q_2 = mq_1 + b$, the relationship between amount of money repaid and time passed is linear. This implies that we should expect that the amount of money that is repaid will increase by equal amounts in equal times.[9]

Relationships Involving Multiple Quantities

Consider the Ideal Gas Equation of State reproduced below.

> p pressure of the gas (Pa)
> V volume of the container (m^3)
> n amount of the gas (mol)
> T temperature of the gas (K)
> R universal gas constant (J/(mol $\cdot$ K))
>
> $pV = nRT$

This equation involves five quantity symbols, one of which is a constant. These are p, V, n, T and the constant R. We will now show how we can use visual cues to relate the value of the various quantities involved in the formula and how this information can help answer certain types of questions.

We begin by working out the relationship between the pressure of the gas, p, and the volume of the container, V. To work out the relationship between two quantities we must of course make sure that the value of the other quantities remain constant. This allows us to tie changes in the value of one quantity to changes in the value of the other quantity.

For the problem above, this implies that the amount of gas, n, and its temperature, T, remain constant. Since R is already a constant, constancy

---

[9]The reader should note the semantic significance behind m and b. Attention to the semantics carried by these quantities should always be noted in linear relationships.

of n and T implies that the term nRT is a constant. We can, therefore, use the symbol k to represent nRT with. This turns the formula above into

$$pV = k$$

which maps onto the conservation form of the equation for inverse proportion, i.e., $q_2 q_1 = k$, with p replacing q_2 and V replacing q_1. The relationship between p and V is, therefore, one of inverse proportion.

We can also solve the formula for one of the quantities and then map the result onto the standard form of the equation for inverse proportion. Following this approach and solving the equation $pV = nRT$ for p yields the following.

$$p = \frac{nRT}{V}$$

Replacing nRT with k as before yields the following formula.

$$p = \frac{k}{V}$$

which maps onto the standard form of the equation for inverse proportion.

It is strongly recommended that the reader should train herself or himself to work with both the standard and conservation forms of these equations. It is the ability to recognize either formulation at a glance that enables one to work out so many relationship in one formula. In the case of the formula $pV = nRT$, one should be able to tell at a glance that p is inversely proportional to V and directly proportional to both n and T, that V is inversely proportional to p and directly proportional to both n and T, that n is directly proportional to both p and V and inversely proportional to T, and that T is directly proportional to both p and V and inversely proportional to n.

Proper training would generate such views of formulas as opposed to *pee vee equals en aar tee!*[10]

As a second example, consider the formula that computes the amount of money that is repaid in financing a purchase with an initial down payment followed by monthly payments.

a_r	amount of maney repaid (\$)
a_d	amount of down payment (\$)
r	monthly payment (\$/month)
t	time passed (months)

$$a_r = rt + a_d$$

---

[10]There are, of course, times when we mindlessly process a formula and see it as *pee vee equals en aar tee* but if that's all that we see ...

To work out the relationship between the amount of money that is repaid, a_r, and the amount of down payment, a_d, we must assume that the monthly payment and time passed are constants. This implies that their product, rt, will also be a constant. Replacing rt with k yields the following. i.e.,

$$a_r = k + a_d$$

which maps onto the standard form of the equation for direct superposition, $q_2 = k + q_1$.

Consider now the relationship between the amount of money that is repaid, a_r, and time passed, t. To work out this relationship, we must set the amount of monthly payment, r, and the down payment, a_d, constant. However, we can *not* turn $r + a_d$ into one constant as the expression is not a valid subexpression of $rt + a_d$. Note that in an expression such as $4 \times 5 + 2$, we would never add 4 and 2.

This implies the need for two constants, one to represent r with and another to represent a_d with. Using m to represent the former and b to represent the latter yields the form of the linear equation.

$$a_r = mt + b$$

so that a_r is a linear function of t.

The two basic guidelines on what can be turned into a constant are the following:

1. A single term or a combination of two or more terms may be turned into a constant.
2. A single factor within a term or a combination of two or more factors within a term may be turned into a constant.[11]

One may never turn a combination of one term and part of another term or parts of two terms into a constant as such a combination has no semantic meaning.

As an example, consider the following expression.

$$q_3 q_2 q_1 + \frac{q_4}{q_5} - q_6$$

As far as the terms are concerned, we can turn any of the terms $q_3 q_2 q_1$, $\frac{q_4}{q_5}$ or q_6 into a constant. Turning $\frac{q_4}{q_5}$ into a constant, as an example, would turn the expression into $q_3 q_2 q_1 + k - q_6$.

---

[11] If a factor involves division, the factors in the dividend and divisor of that division can be considered together or separately.

We may also turn a combination of two or more of these term into a constant. As an example, we can turn the combination $q_3q_2q_1 - q_6$ into a constant, a move that would turn the expression into $k + \frac{q_4}{q_5}$. Turning the combination of all three terms into a constant would turn the whole expression into k.

As far as factors within a term are concerned, we may turn any one or combination of two or more such factors into a constant. As an example, within the first term, we can turn any of the subexpressions q_1, q_2, q_3, q_2q_1, q_3q_1, q_3q_2 or $q_3q_2q_1$ (which represents the whole term) into a constant. Turning q_3q_1 into a constant, as an example, would turn the expression into $kq_2 + \frac{q_4}{q_5} - q_6$.

Within the second term we can turn any of q_4, q_5 or $\frac{q_4}{q_5}$ into a constant.

There is only one factor within the third term and that factor may be turned into a constant.

To establish the relationship between two quantities in a formula we may or may not need to introduce multiple constants. In such cases it is best to first combine as many terms as possible into one constant, and then as many factors as possible within the remaining term into another constant. As an example, if we wish to relate q to q_2 in the equation

$$q = q_3q_2q_1 + \frac{q_4}{q_5} - q_6$$

we first turn the subexpression $\frac{q_4}{q_5} - q_6$ into a constant, b, and then turn the factors q_3q_1 within the first term into another constant, m. This turns the equation into $q = mq_2 + b$ which we recognize as a linear equation.

The ability to recognize relationships at a glance helps us understand more complex relationships between the values of the quantities that appear in a formula. As an example, to increase the value of pressure by a factor of 6 in $pV = nRT$, we can reduce the volume to $\frac{1}{6}$ of its initial size while keeping n and T constant, or we can instead increase the amount of gas by a factor of 6 while keeping V and T constant, or even doubling n and tripling T while keeping V constant.

A wealth of information can be extracted from a formula with relative ease if one knows how to look at them in the manner that we have described above.[12]

---

[12]Here we touch on the basic relationships that can be extracted from a formula but there are others. As an example, the reason the formula $pV = nRT$ has the particular form that it has (as opposed to writing it as $p = \frac{nRT}{V}$ or $\frac{p}{T} = \frac{nR}{V}$ and the like) is that the particular form used, i.e., $pV = nRT$, tells us that work done on or by such a system, whose value, as further studies in physics would show, can be computed through change in the value of the product pV, appears as change in the internal energy of the gas, which, as further studies in physics would show, can be computed through change in the value of nRT. The form $pV = nRT$ is a reminder of this interplay between work done on or by the system and heat added to or removed from the system.

Exercise Set 13

1. List the subexpressions of each of the following expressions below that can be turned into a constant. Rewrite the expression using the constant.

   a. $q_2 q_1 - q_4 q_3$

   b. $q_1 - \dfrac{q_2}{q_3 q_4} - q_6 q_5$

2. Relate the value of the various quantities in each formula by classifying the relationship as *direct superposition, inverse superposition, direct proportion, inverse proportion, linear but not superposition or direct proportion, nonlinear but not inverse proportion.*

   a.

V	voltage across a wire (V)
I	current through the wire (A)
R	resistance of the wire (Ω)

   $$V = IR$$

   b.

T	period of a cyclic event (s)
f	frequency of the cyclic event (1/s)

   $$T = \frac{1}{f}$$

   c.

A	area of triangle (m^2)
b	base of triangle (m)
h	height of triangle (m)

   $$A = \frac{1}{2}bh$$

   d.

p	pressure (Pa)
F	Force (N)
A	area over which force is exerted (m^2)

   $$p = \frac{F}{A}$$

   e.

E	kinetic energy of an object (J)
m	mass of object (m)
v	speed of object (m/s)

   $$E = \frac{1}{2}mv^2$$

f.

m mass of radioisotope left (g)
m_0 initial mass of radioisotope (g)
n number of half-lives passed (1)

$$m = \frac{1}{2^n} m_0$$

g.

p_1 initial pressure (Pa)
p_2 final pressure (Pa)
V_1 initial volume (m^3)
V_2 final volume (m^3)

$$p_1 V_1 = p_2 V_2$$

h.

p_1 initial pressure (Pa)
p_2 final pressure (Pa)
T_1 initial temperature (K)
T_2 final temperature (K)

$$\frac{p_1}{T_1} = \frac{p_2}{T_2}$$

i.

F centripetal force acting on an object (N)
m mass of object (kg)
v speed of object (m/s)
r radius of the circular path (m)

$$F = m\frac{v^2}{r}$$

j.

P perimeter of a rectangle (m)
l length of rectangle (m)
w width of rectangle (m)

$$P = 2(l + w)$$

3. Consider the Ideal Gas Equation of State.

p pressure of the gas (Pa)
V volume of the container (m^3)
n amount of the gas (mol)
T temperature of the gas (K)
R universal gas constant (J/(mol · K))

$$pV = nRT$$

a. Suppose pressure of the gas increases by a factor of 4 while the temperature of the gas increases by a factor of 8. What happens to the volume of the container if the amount of gas remains constant?

b. Describe the change in the temperature of the gas if its volume doubles while the amount of gas decreases by a factor of $\frac{1}{3}$. The pressure remains the same.

c. Describe the change in the volume of the container if the pressure of the gas decreases by a factor of $\frac{1}{3}$, its temperature increases by a factor of 6 and its amount increases by a factor of 2.

d. Describe the change in the amount of gas if the pressure of the gas decreases by a factor of $\frac{1}{3}$ while the volume of the container and the temperature of the gas each increase by a factor 4.

4. Consider the formula for distance covered for motion along a straight line at constant speed.

d final distance from observer (m)
d_0 initial distance from observer (m)
v speed of the object (m/s)
t travel time (s)

$$d = d_0 + vt$$

a. What happens to the final distance between the object and the observer, d, if the initial distance between them, d_0, is increased by 12 m while the covered distance, vt, decreases by 5 m?

b. Suppose the initial distance, d_0, is decreased by 40 m. How would the covered distance, vt, have to change to keep the final distance, d, the same? How can this change be achieved?

c. Suppose that the final distance, d, remains the same while the initial distance, d_0, decreases in such a way to force the covered distance, vt, to double to make up for the loss. How does this affect v_0 if t remains the same?

Appendix A
A Primer on Quantities

Note: The conventions listed in this appendix are those of the International System of Units, SI; the measurement system whose units and conventions are used in the expression of the values of quantities in the sciences by default.[1] The conventions may or may not apply to other measurement systems such as the Imperial System of Units or the US Customary System of Units. In this textbook we limit ourselves to SI for the most part and will, therefore, focus on the features and conventions of this particular measurement system.

A.1 Quantities

A **quantity** is a property of an entity that can be measured objectively. Examples of quantities are length, area, volume, mass, time, temperature, speed, force, energy and the like, but not beauty, kindness or serenity, among others. A listing of some of the more common quantities encountered in everyday life activities and basic sciences is given in Table A.1.1 on the next page.

Entities

A quantity belongs to an entity. At times the entity to which a given quantity belongs is mentioned explicitly as in

The speed of light is approximately 300 000 km/s.

Here the quantity is *speed* and the entity the quantity belongs to is *light*.

---

[1]SI is an improved version of the Metric System of Units.

Quantity Name (Symbol)	Unit Name (Symbol)
cost (c)	dollar ($), pound (£), euro (€), etc.
price (p)	dollar ($), pound (£), euro (€), etc.
length (l)	metre (m), kilometre (km), inch (in), etc.
distance (d)	metre (m), kilometre (km), inch (in), etc.
area (A)	square metre (m^2), square inch (in^2), etc.
volume (V)	cubic metre (m^3), litre (l, ℓ, L), etc.
time (t)	second (s), minute (min), hour (h), etc.
speed (s, v)	metres per second (m/s), etc.
acceleration (a)	metres per square second (m/s^2), etc.
mass (m)	gram (g), atomic mass unit (amu), etc.
weight (W)	newton (N), dyne (dyn), etc.
force (F)	newton (N), dyne (dyn), etc.
pressure (p)	pascal (Pa), atmosphere (atm), etc.
temperature (T, t)	kelvin (K), degree Celsius (°C), etc.
energy (E)	joule (J), calorie (cal), Calorie (Cal), etc.
amount of substance (n)	one (1), mole (mol), etc.
electric current (I, i)	ampere (A), etc.

Table A.1.1: A selection of common quantities and their commonly associated units. The quantity symbol T is used to refer to the thermodynamic temperature, measured using the kelvin as the unit. The quantity symbol t is used to refer to temperature measured using other units, e.g., degree Celsius. The optional quantity symbol for speed, v, relates to the magnitude of the quantity *velocity*. See the discussion ahead on the deviation from SI in classifying the unit *one* as a unit of amount.

At other times the entity that the given quantity belongs to is implied by context. As an example, if the context of discussion is the weather, then the sentence

The temperature is 20 °C.

associates the quantity *temperature* to the *air* at the location where the temperature was measured.

A.2   Quantity Names and Quantity Symbols

To each quantity we associate a name and a symbol.

Quantity names are used to refer to quantities in English sentences and phrases. Examples of quantity names are length, mass, time, and the like. Quantity names are linguistic entities and, as such, are subject to the rules of English grammar. As an example of use of quantity names, we write

To measure the temperature, press and hold the orange button for approximately 3 seconds.

Quantity symbols are used to represent the *values* of quantities in mathematical equations and expressions. Examples of quantity symbols are m, which is used to represent the value of the quantity *mass*; p, which is used to represent the value of the quantity *pressure*, and the like. Quantity symbols are mathematical entities and, as such, are subject to the rules of algebraic grammar. As an example, using t as the quantity symbol to represent the value of the quantity *temperature* in degrees Celsius, we write

$$\frac{9}{5}t + 32$$

to indicate the need to add 32 to $\frac{9}{5}$ of the value of the temperature in degrees Celsius, an activity that converts the value of the temperature in degrees Celsius to the value of the temperature in degrees Fahrenheit.

Quantity symbols are typed using the italic font style. As an example, we write V as the quantity symbol for the value of the quantity *volume* and not V. This helps differentiate between quantity symbols and unit symbols which, as we will soon see, are typed using the upright font style.

Case matters: As we just noted, the symbol used to represent the value of the quantity *volume* is V whereas the lowercase version, v, is used to denote the value of the quantity *speed*.

If necessary, further specification about a quantity is written as a subscript to the quantity symbol. As an example, in a problem where both the value of the mass of an electron and the value of the mass of a proton are being discussed, we can represent the former as m_e and the latter as m_p. As another example, if values of energies of kinetic and potential varieties are under investigation, we may choose to represent the former as E_k and the latter as E_p. Note that *in all cases the main letter is the quantity symbol and the subscript acts as a qualifier.* Therefore, we write m_e (and not e_m) to refer to the value of the mass of an electron and we write E_k (and not K_e) to refer to the value of kinetic energy.[2]

Quantity symbols represent values and are not abbreviations for the names of quantities that are subject to the rules of English grammar.

A.3   Units

Units are used to measure the values of quantities. Examples of units are metre, gram, degree Celsius, and the like.

Many different units can be used to measure the value of a given quantity. As an example, we can use the units metre, kilometre, inch, foot and others to measure the value of the quantity *length* with. As another example, we can use the units degree Celsius, degree Fahrenheit, kelvin and others to measure the value of the quantity *temperature* with.

A.4   Unit Names and Unit Symbols

To each unit we associate a name and a symbol.

Unit names are used to refer to units in English sentences and phrases. Examples of unit names are metre, gram, second, and the like. Unit names are linguistic entities and, as such, are subject to the rules of English grammar. As an example of use of unit names, we write

> To measure the temperature, press and hold the orange button for approximately 3 seconds.

Unit names begin with lowercase letters unless they appear at the beginning of a sentence or in titles where the first letters in the words in the title are

---

[2]This last convention is followed by some but not all. Physicists tend to follow this convention closely. Others, notably those active in branches of chemistry and the biological sciences, on the other hand, frequently break this convention. As an example, it is common to see K_e, or even worse, KE, or some such similar representations of the value of kinetic energy in chemistry or biology textbooks.

capitalized.[3] This helps differentiate between the unit and the individual when the unit name is fashioned after the name of an individual. As an example, the word *kelvin* refers to the unit of temperature whereas *Kelvin* refers to the individual whose name was used in the naming of this unit.

Unit symbols are used to represent the *sizes* of units in mathematical equations and expressions. Examples of unit symbols are the symbol m used to represent the size of the unit *metre*, kg used to represent the size of the unit *kilogram*, and the like. Unit symbols are mathematical entities and, as such, are subject to the rules of algebraic grammar. As an example, using kg as the unit symbol to represent the size of the unit *kilogram*, we write

$$m = 20.5 \text{ kg}$$

to indicate that the mass has a value of 20.5 kilograms.

Unit symbols are typed using the upright font style. As an example, we write m to refer to the size of the unit *metre*, not m. This helps differentiate between unit symbols and quantity symbols which, as we saw earlier, are typed using the italic font style.

Case matters.

SI prefixes can be used along with units to generate larger or smaller units. Examples of SI prefixes are *kilo, centi, milli* and the like. The prefix names and symbols are used along with the names and symbols of the various units without intervening white spaces: We write kilogram and kg, and not kilo gram or k g. The casing of these prefixes matters. As an example, the symbol mg is the used to represent the size of the unit milligram which is equal to one thousandth of a gram, whereas the symbol Mg is used to represent the size of the unit megagram which is equal to one million grams.

Mixed prefixes should never be used.

Unit symbols represent sizes and are not abbreviations for the names that are subject to the rules of English grammar. As an example, the period should not be used at the end of a unit symbol unless is appears at the end of a sentence. As another example, the plural *s* should not be used along with unit symbols: The representation ms is not interpreted as the plural form of metre but as the unit millisecond. As a third example, the hyphen should not be used even if the unit symbol acts as an adjective, e.g., we write 'a 2.5 kg object' and not 'a 2.5-kg object'.[4]

---

[3]We remind the reader that the conventions listed in this appendix are those of SI. Other measurement systems may not follow these conventions.

[4]But we do write *a 2.5-kilogram object.*

A.5   The Value of a Quantity

In the sentence

The distance from Toronto to Montreal is 542 km.

we refer to 542 km as the value of the quantity *distance*. The **value of a quantity** is made up of two parts. The first is the numerical part (e.g., the 542 by itself in the example above), called the **numerical value of the quantity**. This is followed by an empty space, implying multiplication, and the symbol representing the size of the unit used in measuring the value of the quantity.

Examples

In each case identify the quantity being measured (list both the quantity name and quantity symbol), the entity to which the quantity belongs, the value of the quantity, the numerical value of the quantity, and the unit used (list both the unit name and unit symbol).

1. The car has a mass of 1243 kg.
2. The reaction generated 5.2 mol of oxygen.
3. There are 760 frogs in the sample.
4. The fridge uses 970 W of power.

Answers

1. Quantity name: mass
   Quantity symbol: m
   Entity: car
   Value of the quantity: 1243 kg
   Numerical value of the quantity: 1243
   Unit name: kilogram
   Unit symbol: kg

2. Quantity name: amount
   Quantity symbol: n
   Entity: oxygen
   Value of the quantity: 5.2 mol
   Numerical value of the quantity: 5.2
   Unit name: mole
   Unit symbol: mol

3. Quantity name: amount
   Quantity symbol: n
   Entity: frogs
   Value of the quantity:[5] 760 1
   Numerical value of the quantity: 760
   Unit name: one
   Unit symbol: 1

4. Quantity name: power
   Quantity symbol: P
   Entity: fridge
   Value of the quantity: 970 W
   Numerical value of the quantity: 970
   Unit name: watt
   Unit symbol: W

Exercise Set A

In each case identify the quantity being measured (list both the quantity name and quantity symbol), the entity to which the quantity belongs, the value of the quantity, the numerical value of the quantity, and the unit used (list both the unit name and unit symbol).

1. The reaction generated 45.5 g of CO_2.
2. The book costs \$9.20.
3. Rate of discount is 15%.
4. The train was travelling at a speed of 200 km/h.
5. The lot has an area of 4000 ft^2.
6. The operation will take 1.5 h.
7. The temperature of the surface of the sun is 6000 °C.
8. The bullet has a kinetic energy of 2025 J.
9. This BMW can accelerate at 12.66 m/s^2.
10. Charlie[6] has a mass of 32 kg.

---

[5]The notation 760 1 implies seven hundred sixty *ones*. This is similar to 760 g or 760 km, etc., with 1 acting as the unit symbol for *one*, a unit of counting.

The use of 1 as unit symbol may, unfortunately, lead to interpretation errors with the reader assuming that the writer might have meant to say 7601, i.e., seven thousand six hundred one as opposed to 760 1, i.e., seven hundred sixty *ones*. For this reason it is common to replace the unit symbol for one with a word that describes the nature of the one, e.g., 760 frogs with the understanding that frogs is used, not as a name, but a symbol to represent the size of the unit of counting.

For the most part in this textbook we will remain formal and use 1 as the unit symbol for counting but in practice we recommend the use of a word that describes the nature of the one. Readers insisting on using unit symbols may want to consider placing a word descriptive of the nature of the one as a subscript to the unit symbol, e.g., 1_{frog}.

[6]A dog in my neighborhood :)

Appendix B
A Primer on Chemistry

B.1   Atoms

Atoms are tiny building blocks that group in certain ways to generate the materials of this world. Over a hundred different atoms have been detected or generated so far.

Each atom is given a **chemical name** and a **chemical symbol**. A chemical symbol is made up of either a single uppercase letter which maps onto the initial letter in the name of the atom in Latin, or two letters: An uppercase letter which maps onto the initial letter in the name of the atom in Latin, followed by a lowercase letter which corresponds to some other letter in the name of the atom in Latin.[1] As examples we have the hydrogen atom and its chemical symbol H, the helium atom and its chemical symbol He, and so on.[2,3]

---

[1] The choice of a single language to base names and chemical symbols on imposes uniformity which in turn aids communication. The choice of Latin, an ancient language, as the language to base names and chemical symbols on reflects the fact that, during the Middle Ages in Europe, Latin was the medium in which scholarly ideas were exchanged and, as such, the language had already been familiar to many. In addition, Latin is at the root of many Western languages and, as such, its choice was seen as being neutral albeit not globally so.

It should also be noted that despite the agreement on the choice of Latin as the official language to base names and chemical symbols on, in many countries language-specific names for one or more atoms persist. As an example, we use the word *iron* in English rather than its Latin counterpart *ferrum* on which the chemical symbol for iron, Fe, is based. Whether or not the Latin name or some language-specific name is used to refer to a given atom, the corresponding chemical symbol must *always* be based on the letters in the name of that atom in Latin as specified.

[2] Chemical symbols are *not* to be seen as abbreviations of the names of atoms (which would make them subject to the rules of grammar), but universal symbols that can be used to represent atoms with. We were, of course, smart in our choice of symbols and, rather than inventing a host of new symbols to refer to the many atoms, we chose to use symbols that appear as letters in the writings of their names so that we can easily associate names and chemical symbols.

[3] Casing is important: Co is an atom of cobalt whereas CO is carbon monoxide, a

Atoms are organized in a 2×2 table with atoms that exhibit similar chemical behaviours grouped into columns and those that naturally follow each other placed next to one another along the rows from left to right, or from the end of one row to the start of the next. The resulting table is called the *Periodic Table of the Elements* often shortened to *the Periodic Table*. The Periodic Table is shown in Figure B.2.2.

Exercise Set B.1

1. What is an atom?
2. Using the Periodic Table in Figure B.2.2 determine the chemical symbol that is associated with each of the following atoms.

   a. oxygen   c. argon   e. zinc   g. carbon
   b. magnesium   d. sulfur   f. chlorine   h. calcium

3. Using the Periodic Table in Figure B.2.2 name the atom that is represented by each of the following chemical symbols.

   a. N   c. F   e. Fe   g. C
   b. Ne   d. Na   f. Pb   h. I

4. What is common to atoms within the same column in the Periodic Table?
5. Using the Periodic Table in Figure B.2.2 name the atom that follows the one named in each case below.

   a. lithium   c. nitrogen   e. argon   g. silver
   b. magnesium   d. fluorine   f. hydrogen   h. helium

B.2   Subatomic Particles

An atom is composed of a number of smaller particles, called **subatomic particles**. Of the many different subatomic particles, three play a major role in chemistry:[4] The **proton**, with a mass of 1 amu and an electric charge of 1 e, the **neutron**, with a mass of roughly 1 amu and an electric charge of 0 e, and the **electron**, with a mass of roughly 0.000 549 amu (about $\frac{1}{2000}$ of the mass of

---

molecule (a grouping of atoms) made up of 1 carbon atom and 1 oxygen atom.

[4]In what follows the unit symbol amu relates to the unit name *atomic mass unit* which is a tiny unit of mass suitable for measuring the masses of subatomic particles, atoms, and molecules, and the unit symbol e relates to the unit name *elementary charge* which is a tiny unit of electric charge suitable for measuring the electric charges of subatomic particles, ions, and polyatomic ions.

a proton or a neutron) and an electric charge of -1 e. Protons and electrons exert attractive electromagnetic forces on each other whereas protons and protons, as well as electrons and electrons exert repulsive electromagnetic forces on each other. Neutrons are not subject to electromagnetic forces.

Protons and neutrons reside in the nuclei (singular: **nucleus**) of atoms: tiny volumes at the centre of atoms within which they are packed. Electrons travel around the nuclei at distances that are large compared to the sizes of the nuclei. As a rough analogy, if the nucleus of an atom had the size of a tennis ball, the electrons would be scattered over an area the size of a soccer field.

Protons

The number of protons in the nucleus of an atom sets the type of the atom. As an example, any atom with 6 protons in its nucleus is a carbon atom.

We refer to the number of protons in the nucleus of an atom as its **atomic number**. Since the atomic number of an atom sets the type of the atom, sequencing the many atoms according to their atomic number seems to be a logical choice. This means that the numerical position of an atom in the sequence maps onto the atom's atomic number. As an example, the atomic number of carbon is 6 which means that a carbon atom has 6 protons in its nucleus. It also means that carbon is the 6th atom in the order of listing of atoms.

In the Periodic Table, the atomic number is written on top of the chemical symbol for the atom within the cell for that atom. In chemical equations, the atomic number is written as a subscript on the bottom, left side of the chemical symbol for the atom. Figure B.2.1 below illustrates the placement of the atomic number of carbon in the Periodic Table and in chemical equations.

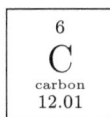

| 6 |
| C |
| carbon |
| 12.01 |

$_6\text{C}$

Placement of the atomic number
of carbon in the Periodic Table

Placement of the atomic number
of carbon in chemical equations

Figure B.2.1: Placement of the atomic number of carbon in the Periodic Table and in chemical equations

1																	18
1 H hydrogen 1.008A	2											13	14	15	16	17	2 He helium 4.003
3 Li lithium 6.968A	4 Be beryllium 9.012											5 B boron 10.82A	6 C carbon 12.01A	7 N nitrogen 14.01A	8 O oxygen 16.00A	9 F fluorine 19.00	10 Ne neon 20.18
11 Na sodium 22.99	12 Mg magnesium 24.31A	3	4	5	6	7	8	9	10	11	12	13 Al aluminium 26.98	14 Si silicon 28.09A	15 P phosphorus 30.97	16 S sulfur 32.07A	17 Cl chlorine 35.45A	18 Ar argon 39.95
19 K potassium 39.10	20 Ca calcium 40.08	21 Sc scandium 44.96	22 Ti titanium 47.87	23 V vanadium 50.94	24 Cr chromium 52.00	25 Mn manganese 54.94	26 Fe iron 55.85	27 Co cobalt 58.93	28 Ni nickel 58.69	29 Cu copper 63.55	30 Zn zinc 65.38	31 Ga gallium 69.72	32 Ge germanium 72.63	33 As arsenic 74.92	34 Se selenium 78.96	35 Br bromine 79.91A	36 Kr krypton 83.80
37 Rb rubidium 85.47	38 Sr strontium 87.62	39 Y yttrium 88.91	40 Zr zirconium 91.22	41 Nb niobium 92.91	42 Mo molybdenum 95.96	43 Tc technetium	44 Ru ruthenium 101.1	45 Rh rhodium 102.9	46 Pd palladium 106.4	47 Ag silver 107.9	48 Cd cadmium 112.4	49 In indium 114.8	50 Sn tin 118.7	51 Sb antimony 121.8	52 Te tellurium 127.6	53 I iodine 126.9	54 Xe xenon 131.3
55 Cs caesium 132.9	56 Ba barium 137.3	lanthanoids	72 Hf hafnium 178.5	73 Ta tantalum 180.9	74 W tungsten 183.8	75 Re rhenium 186.2	76 Os osmium 190.2	77 Ir iridium 192.2	78 Pt platinum 195.1	79 Au gold 197.0	80 Hg mercury 200.6	81 Tl thallium 204.4A	82 Pb lead 207.2	83 Bi bismuth 209.0	84 Po polonium	85 At astatine	86 Rn radon
87 Fr francium	88 Ra radium	actinoids	104 Rf rutherfordium	105 Db dubnium	106 Sg seaborgium	107 Bh bohrium	108 Hs hassium	109 Mt meitnerium	110 Ds darmstadtium	111 Rg roentgenium	112 Cn copernicium		114 Fl flerovium		116 Lv livermorium		

lanthanoids	57 La lanthanum 138.9	58 Ce cerium 140.1	59 Pr praseodymium 140.9	60 Nd neodymium 144.2	61 Pm promethium	62 Sm samarium 150.4	63 Eu europium 152.0	64 Gd gadolinium 157.3	65 Tb terbium 158.9	66 Dy dysprosium 162.5	67 Ho holmium 164.9	68 Er erbium 167.3	69 Tm thulium 168.9	70 Yb ytterbium 173.1	71 Lu lutetium 176.0
actinoids	89 Ac actinium	90 Th thorium 232.0	91 Pa protactinium 231.0	92 U uranium 238.0	93 Np neptunium	94 Pu plutonium	95 Am americium	96 Cm curium	97 Bk berkelium	98 Cf californium	99 Es einsteinium	100 Fm fermium	101 Md mendelevium	102 No nobelium	103 Lr lawrencium

Figure B.2.2: The Periodic Table of the Elements (Adapted from the publication *the International Union of Pure and Applied Chemistry (IUPAC) Periodic Table of the Elements, May 2013*. The phrase *atomic weight* has been replaced with *atomic mass*. Uncertainties in the last digit in atomic masses are omitted. The letter 'A' appearing after an atomic mass indicates that the atomic mass interval is replaced by the average value of the endpoints of that interval. The number of significant digits were retained in the process. As in the original document, we have refrained from quoting atomic masses when the element lacks isotopes with characteristic isotopic abundance in natural terrestrial samples. Alternative spellings for aluminium and caesium are aluminum and cesium.)

Neutrons

Atoms of the same kind, i.e., those with the same number of protons, may have different number of neutrons. We refer to atoms of the same kind that have different number of neutrons as **isotopes** of that atom.

We refer to the total number of protons and neutrons in an atom as its **mass number** and we write the mass number as a superscript on the top, left side of the chemical symbol for the atom. As an example, we write ^{13}C to refer to the isotope of carbon that has 13 protons and neutrons.[5,6,7]

---

[5]Two other notations in common use are C-13 and carbon-13. Unlike the notation ^{13}C, the notations C-13 and carbon-13 are both supported in text-based editors. The latter is geared at audiences who may not necessarily know what atom the chemical symbol represents.

[6]The use of the adjective *mass* in *mass number* arises from the fact that the numerical value of the total *number of* protons and neutrons in the nucleus of an isotope is equal to the numerical value of the *mass* of that isotope. As an example, ^{13}C has a mass number of 13, meaning that it has a total of 13 protons and neutrons. Since the mass of each proton is 1 amu and the mass of each neutron is approximately 1 amu, the mass of this particular isotope of carbon can be computed as

1 amu/particle $\times$ 13 particles $=$ 13 amu

with the word particle used to refer to protons and neutrons (the masses of any other subatomic particles are either 0 or negligible). This computation shows that the numerical value of the total *number* of protons and neutrons in the nucleus of a ^{13}C isotope is equal to the numerical value of the *mass* of a ^{13}C isotope in amu. The following diagram illustrates the difference between *mass number* and *mass* further.

different quantities

mass number   mass

1 amu/particle $\times$ 13 particles $=$ 13 amu

equal numerical
values

The equality between the numerical value of the total *number* of protons and neutrons in the nucleus of an isotope and the numerical value of its *mass* prompts us to refer to the total number of the protons and neutrons as *mass* number to remind ourselves that a knowledge of this number allows us to determine the mass of that isotope by keeping the numerical value and changing its unit from one (a unit of counting) to amu (a unit of mass).

[7]The number of neutrons in a given isotope can be computed by subtracting the number of protons in the corresponding atom, i.e., the atomic number, from the total number of protons and neutrons in that isotope, i.e., its mass number. As an example, since a carbon

In theory, one can associate any number of neutrons to a given type of atom, starting with no neutrons, and moving on to 1 neutron, 2 neutrons, 3 neutrons, and so on. In practice, however, we find that only some of the possible isotopes of a given atom are stable and that the number of neutrons in stable isotopes of an atom is usually either equal, or else close, to the number of protons in their nuclei. Unstable isotopes change into other entities through processes that result in the splitting of their nuclei or rearrangement of the particles within their nuclei. Such nuclear changes are often accompanied by **radiation** which consists of particles being thrown out of the nucleus at extremely high speeds. We refer to isotopes that exhibit this kind of behaviour as **radioisotopes**. A radioisotope that has gone through nuclear changes is said to have **decayed**.

To illustrate matters, let us take a look at the isotopes of carbon. Carbon has a number of known isotopes of which only two are stable: The first, found some 99% of the time in nature, is ^{12}C, the isotope of carbon with 6 neutrons and the second, found some 1% of the time in nature, is ^{13}C, the isotope of carbon with 7 neutrons. All other isotopes of carbon are unstable and decay by emitting radiation. The most important of the radioisotopes of carbon, ^{14}C, the isotope of carbon with 8 neutrons used in dating old objects, is found in nature in negligible amounts. The particular type of radiation that a ^{14}C isotope goes through results in the conversion of a neutron in its nucleus into a proton and an electron. The proton, which stays in the nucleus, increases the atomic number of carbon from 6 to 7, resulting in the conversion of carbon to nitrogen. The radiation consists of the electron, thrown out of the nucleus at a speed close to the speed of light. The departure of the electron leaves the nitrogen with an extra positive charge. The **nuclear equation** for ^{14}C is

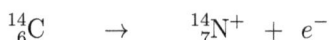

$$^{14}_{6}C \quad \rightarrow \quad ^{14}_{7}N^{+} + e^{-}$$

Note that the mass number in this particular type of decay does not change. This is because the loss of the neutron is accompanied by the generation of a proton so that the mass number (which is the total number of protons and neutrons) stays the same.[8]

Note that the masses of different isotopes of an atom can be noticeably different: ^{12}C has a mass of 12 amu, ^{13}C has a mass of 13 amu, and so on. We can *not*, therefore, say that the mass of *a* carbon atom is 12 amu or 13 amu, etc., as it depends on which particular isotope of carbon we are dealing

---

atom has 6 protons, a ^{13}C isotope must have $13 - 6$ or 7 neutrons.

[8]The type of radiation that a ^{14}C isotope goes through is also experienced by certain isotopes of some other atoms but this type of radiation is not the only type of radiation there is. There are radioisotopes that, on their way to stability, throw out a grouping of two protons and two neutrons. There are radioisotopes that, on their way to stability, throw out a photon. And there are others.

with. This dependence of the mass of an isotope on its type is unwelcome as it complicates calculations that are based on masses of atoms.

What we can do to circumvent this problem, and a solution that seems to work well in practice, is to compute the average mass of all the different isotopes of an atom in a given sample of that atom and consider this average mass to be the mass of each and every atom of that kind in that sample, regardless of which isotope each actually is. If the sample comes from nature, we refer to the resulting average mass as the **atomic mass** of that atom. In the Periodic Table the atomic mass of an atom is written underneath the chemical name of that atom. As an example, the cell for the carbon atom in the Periodic Table reproduced in Figure B.2.3 below

6

C

carbon
12.01

The atomic mass of carbon is
12.01 amu/atom

Figure B.2.3: Placement of the atomic mass in the
Periodic Table

tells us that the average mass of a carbon atom from a sample from nature (which contains 99% ^{12}C, 1% ^{13}C, and negligible amounts of ^{14}C) is 12.01 amu/atom.

Electrons

In an atom the number of electrons moving around the nucleus is equal to the number of protons in its nucleus. Thus a carbon atom has 6 electrons, the same as the number of protons in its nucleus.

An atom with an unbalanced charge is referred to as an **ion**. Such unbalanced charges normally appear when atoms gain or lose one or more electrons.[9] An atom that has gained one or more electrons and is, therefore,

---

[9] In theory, it is possible for an ion to form by gaining or losing protons. In practice, however, such events are rare. To force a proton out of a nucleus, one must overcome the effect of strong nuclear forces that hold the nucleus together. At close range, strong

carrying a net negative charge, is called an **anion** and one that has lost one or more electrons and is, therefore, carrying a net positive charge, is called a **cation**. We write F^- to refer to the ion that forms when a fluorine atom gains an extra electron and we write S^{2-} to refer to the ion that forms when a sulfur atom gains two extra electrons, and so on. In addition, we write Na^+ to refer to the ion that forms when a sodium atom loses an electron (which implies that the ion is carrying an extra *positive* charge) and we write Ca^{2+} to refer to the ion that forms when a calcium atom loses 2 electrons (which implies that the ion is carrying two extra *positive* charges), and so on.

It is also possible for a grouping of atoms to carry an unbalanced charge. Such entities are referred to as **polyatomic ions** with the prefix *poly* meaning *many*. An example of a polyatomic ion is NO_3^{2-}. This is a grouping of 1 nitrogen atom and 3 oxygen atoms that carries two extra electrons compared to the sum of the number of protons in the nuclei of the atoms involved. Please note that the notation makes no reference whatsoever as to the *location* of the extra electrons.

Exercise Set B.2

1. Name the three subatomic particles that play a major role in chemistry.

2. What is the mass of

   a. a proton?   b. a neutron?   c. an electron?

3. What is the electric charge of

   a. a proton?   b. a neutron?   c. an electron?

4. What is the *nucleus* of an atom?

5. What is *atomic number*?

6. What information do we get from the number of protons in the nucleus of an atom?

7. Where can one find the atomic number within the cell of an atom in the Periodic Table?

8. Where do we write the atomic number of an atom when we represent the atom as a symbol in chemical equations?

---

nuclear forces dominate and easily overcome the effects of other forces such as the repulsive force between protons, pushing them apart. To push a proton *into* a nucleus, on the other hand, one must overcome the strong repulsive force between the positively charged nucleus and the proton being pushed in. Such repulsive forces dominate at larger distances and easily overcome the strong force that is only effective in keeping the nucleus together when distances between particles are small.

9. Using the Periodic Table in Figure B.2.2 determine the atomic number of each of the following atoms:

   a. Na   c. Br   e. B   g. O
   b. Au   d. N   f. He   h. S

10. What are isotopes?

11. What is mass number?

12. Explain the use of the adjective *mass* in the quantity name *mass number.*

13. Where do we write the mass number of an atom when we represent the atom as a symbol in chemical equations?

14. Write the mass number and the mass of each of the following isotopes.

   a. $^{17}_{8}O$ b. $^{35}_{16}S$ c. $^{99}_{43}Tc$ d. $^{131}_{53}I$

15. How do the number of neutrons compare to the number of protons in stable isotopes of an atom?

16. Write each of the following using two other notations

   a. ^{131}I b. S-35   c. hydrogen-3

17. How many neutrons are there in each of the following isotopes? Use the Periodic Table in Figure B.2.2 to find the number of protons in the corresponding atoms.

   a. ^{17}O c. ^{99}Tc e. ^{19}F g. ^{186}U
   b. ^{35}S d. ^{131}I f. ^{3}H h. ^{31}P

18. An **alpha radioisotope** is a radioisotope that decays by emitting a group of 2 protons and 2 neutrons as a unit called an **alpha particle**. An example of an alpha radioisotope is americium-241. We can use the information above on the behaviour of an alpha radioisotope to write the nuclear equation for ^{241}Am. First, we note that americium-241 will throw out a complex of 2 protons and 2 neutrons. This complex corresponds to the nucleus of an ^{4}He atom: The nucleus of an ^{4}He atom has 2 protons (See the Periodic Table) and a total of 4 protons and neutrons which means that it has 2 neutrons. An alpha particle, then, is identical to an $^{4}_{2}He$ atom without its 2 electrons, i.e., it is the helium ion, $^{4}_{2}He^{2+}$. We write:

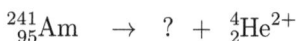

$$^{241}_{95}Am \rightarrow ? + ^{4}_{2}He^{2+}$$

To identify the new atom, we note that the departure of 2 protons from the nucleus of the americium atom lowers the atomic number of the americium atom by 2, from 95 to 93. The americium atom, then, changes to a neptunium atom, Np. Furthermore, the departure of 4 protons and neutrons from the nucleus of the americium atom lowers the mass number of the americium atom by 4, from 241 to 237. The nuclear equation for americium-241 is[10]

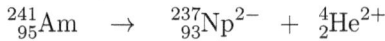

$$^{241}_{95}\text{Am} \quad \rightarrow \quad ^{237}_{93}\text{Np}^{2-} + {}^{4}_{2}\text{He}^{2+}$$

Each of the following is an alpha radioisotope. In each case, following the line of reasoning given above, write the nuclear equation for the radioisotope.

a. ^{198}Ra b. ^{238}U c. ^{211}At

19. A **beta radioisotope** is a radioisotope that decays by emitting an electron, referred to as a **beta particle**, from its nucleus. The electron emerges as a result of the conversion of a neutron in the nucleus into a proton and an electron. An example of a beta radioisotope is carbon-14 whose nuclear equation was given in the body of the text.[11]

Each of the following is a beta radioisotope. In each case, following the argument given for carbon-14 in the body of the text, write the nuclear equation for the radioisotope.

a. ^{33}P b. sulfur-35   c. H-3

20. A **gamma radioisotope** is a radioisotope that decays by emitting a photon, referred to as a **gamma particle**, from its nucleus. Such radioisotopes are said to be **metastable**, a state that is noted by writing the lowercase letter 'm' after the mass number of the radioisotope. An example of a gamma radioisotope is technetium-99m. The nuclear equation for technetium-99m is

$$^{99m}_{43}\text{Tc} \quad \rightarrow \quad ^{99}_{43}\text{Tc} + \gamma$$

Note that the letter 'm' after the mass number has been removed. The symbol γ, the lowercase Greek letter, gamma, is used to represent the gamma particle.

---

[10] Also written as

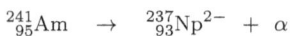

$$^{241}_{95}\text{Am} \quad \rightarrow \quad ^{237}_{93}\text{Np}^{2-} + \alpha$$

where the symbol α, the lowercase Greek letter, alpha, is used to represent the alpha particle.

[11] The nuclear equation for C-14 given in the body of the text could also be written as

$$^{14}_{6}\text{C} \quad \rightarrow \quad ^{14}_{7}\text{N}^{+} + \beta$$

where the symbol β, the lowercase Greek letter, beta, is used to represent the beta particle.

Each of the following is a gamma radioisotope. In each case, following the nuclear reaction for technetium-99m as a guide, write the nuclear equation for the radioisotope.

a. ^{60m}Co   b. ^{65m}Zn   c. ^{137m}Cs

21. What is atomic mass?

22. Using the Periodic Table in Figure B.2.2 determine the atomic mass of each of the following atoms.

a. O   c. Tc   e. F   g. U
b. S   d. I   f. H   h. P

23. How does the number of electrons compare to the number of protons in an atom?

24. How many electrons are there in each of the following atoms? Use the Periodic Table in Figure B.2.2 to get the number of protons in the atoms.

a. O   c. Tc   e. F   g. U
b. S   d. I   f. H   h. P

25. What is an ion? An anion? A cation?

26. a. A fluorine ion has 1 extra electron compared to a fluorine atom. Write its chemical symbol.
    b. An oxygen ion has 2 extra electrons compared to an oxygen atom. Write its chemical symbol.
    c. A lithium ion has 1 fewer electron compared to a lithium atom. Write its chemical symbol.
    d. An aluminum ion has 3 fewer electrons compared to an aluminum atom. Write its chemical symbol.

27. How does the number of electrons compare to the number of protons in an ion?

B.3   Molecules and Ionic Substances

By the term **molecule** we mean a grouping of atoms that is distinct from other atoms or groupings of atoms. As an example, a molecule of carbon dioxide is made up of a grouping of one carbon atom and two oxygen atoms. We write this as CO_2 which is short for C_1O_2.[12] A molecule of CO_2 is separate and distinct from other atoms and molecules including other CO_2 molecules.

---

[12]Subscripts are used to specify the number of atoms of each kind that are present in a molecule with subscript 1 omitted by convention.

Molecules form when atoms with tendencies to attract electrons meet. This tendency is rooted in the tendency to maximize stability and can be achieved, at least partially, if the atoms involved *share* one or more of their outer electrons. The shared electrons spend most of their time within the space between the atoms. We refer to the association thus formed as a **covalent bond**.

If the two interacting atoms have the same tendencies to attract electrons or the difference between their tendencies to attract electrons is small, then they share their electrons equally or almost equally. This results in the formation of a bond that does not exhibit much polarity and is, for this reason, called a **nonpolar covalent bond**. An example of a nonpolar covalent bond is the bond between the two atoms of hydrogen in an H_2 molecule.

If, on the other hand, the difference between the tendencies of the two atoms to attract electrons is pronounced, then the atom with the higher tendency to attract electrons will pull the shared electrons closer to itself, resulting in its side becoming slightly negative[13] and the other side becoming slightly positive[14]. Such bonds do exhibit polarity and are, for this reason, called **polar covalent bonds**. An example of a polar covalent bond is the bond between an atom of hydrogen and an atom of oxygen. In such a bond, the much higher tendency of the oxygen atom to attract electrons implies that the oxygen atom will pull the shared electrons towards itself, resulting in the O side to become $\delta-$ and the H side to become $\delta+$.

While some atoms achieve a more stable state by gaining extra electrons, others do so by *losing* electrons. If an atom with a tendency to lose electrons nears another which has a tendency to gain electrons, then the resulting tendencies force the transfer of one or more electrons from the atom that has a tendency to lose electrons to the one that has a tendency to gain electrons, generating positive and negative ions. The force of attraction between the resulting oppositely charged ions is large, resulting in these ions to stick, not just to each other, but also to any other ions that may be present around them.[15] In nature we tend to find repetitive cycles of such ions in crystal form. We refer to such substances as **ionic substances**.

An example of an ionic substance is table salt which is generated from repetitive cycles of Na^+ ions and Cl^- ions in three dimensions. Sodium

---

[13]We denote this as $\delta-$ with the lowercase Greek letter δ (read *delta*) standing for the phrase *slightly*.

[14]We denote this as $\delta+$ with the lowercase Greek letter δ (read *delta*) standing for the phrase *slightly*.

[15]The force of attraction of an ion on other ions surrounding it does not diminish if more ions are introduced in its vicinity. If an electron exerts a certain attractive force on a proton at a certain distance to it on its left, the electron will exert the same force on another proton that may be placed at the same distance from it in any other direction. It is *not* the case that the presence of the second proton forces the electron to divide the force that it exerted on the one proton between the two when the second proton was introduced.

atoms have a strong tendency to lose an electron and chlorine atoms have a strong tendency to gain an electron. If a sodium atom gets close to a chlorine atom, their tendencies result in the transfer of an electron from the sodium atom to the chlorine atom, generating a sodium ion, Na^+, and a chlorine ion, Cl^-. The charged ions are strongly attracted to each other, forming an association between the ions called an **ionic bond**. If there are many Na^+ ions and Cl^- ions, they stick together in a three-dimensional crystal-like structure which we see as grains of salt. The ions that exist within the same crystal are connected to each other. Those that exist in separate crystals are not.

To refer to an ionic substance we list the types and amounts of the atoms that appear as ions in a repeating pattern within that ionic substance. As an example, we refer to table salt as NaCl, keeping in mind that it is Na^+ ions and Cl^- ions that are found within table salt. We also understand that the notation does not imply that we have a molecule (which is an entity that is distinct from other entities) but that there is a repeating cycle of Na^+ ions and Cl^- ions that is being dealt with. Since the charges are often not written in representing the repeating cycles of ions in ionic substances, the notation relies on an a priori knowledge on the part of the reader as to whether the representation refers to a molecule or an ionic substance. As an example, it is expected that the reader knows that the notation H_2O refers to a molecule with a covalent bond between the oxygen atom and each of the hydrogen atoms whereas the notation NaCl represents a repeating pattern of ionic bonds between Na^+ ions and Cl^- ions.[16]

Polyatomic ions may also be involved in forming ionic bonds with other ions or polyatomic ions. The notation $Ca(NO_3)_2$, as an example, denotes an ionic bond between 1 Ca^{2+} ion and 2 NO_3^- ions.

As we noted above, we use the notation CO_2 to refer to a single molecule consisting of 1 carbon atom and 2 oxygen atoms and we use the notation $MgCl_2$ to refer to a repetitive cycle of 1 Mg^{2+} ion and 2 Cl^- ions in the ionic substance. If we wish to refer to a number of such molecules or repetitive cycles, then we place a number on the left side of the formula for that molecule or repetitive cycle. As an example, we write $8CO_2$ to refer to 8 molecules of CO_2. Since each molecule of CO_2 has 1 C atom, the total number of C atoms in $8CO_2$ molecules is 8×1 or 8. Similarly, since each molecule of CO_2 contains 2 O atoms, the total number of O atoms in 8 molecules of CO_2 is

---

[16]One way to tell whether we are dealing with molecules or ionic substances is to learn how to classify atoms as **metals**, which are atoms that tend to lose electrons, and **nonmetals**, which are atoms that tend to gain electrons. Metals are found on the left side and the middle of the Periodic Table while nonmetals are found on the right side of the Periodic Table. A gray area does exist, occupied by atoms that are referred to as **metalloids**. Covalent bonds are formed between nonmetals whereas ionic bonds are formed between metals and nonmetals. We will not delve into such matters in this textbook and refer the reader to any standard textbook on fundamentals of chemistry for more detail.

8×2 or 16. Given $8CO_2$, then, we can multiply the number 8 by the subscript on each atom in turn to calculate the total number of that type of atom in a collection of 8 molecules of CO_2. The same remarks hold true when working with representations of ionic substances.

Exercise Set B.3

1. What is a covalent bond? A nonpolar covalent bond? A polar covalent bond?

2. What is a molecule?

3. How many atoms of each kind are there in each of the following?

   a. CH_4 c. SO_3 e. $C_6H_{12}O_6$ g. Cl_2
   b. C_2H_6 d. N_2O_4 f. O_3 h. CS_2

4. What is an ionic bond?

5. What is an ionic substance?

6. What ions are present in each of the following ionic substances?

   a. $MgCl_2$ b. NaI c. Na_2S

7. What are metals? Nonmetals? Metalloids?

8. How many atoms of each kind are there in each of the following?

   a. $5CH_4$ c. $8SO_3$ e. $15C_6H_{12}O_6$ g. $3Cl_2$
   b. $2C_2H_6$ d. $12N_2O_4$ f. $2O_3$ h. $95CS_2$

B.4   Chemical Reactions

A **chemical reaction** may be thought of as a process which results in changes in the composition of at least one molecule.[17]

---

[17]Processes that leave molecular compositions and atomic identities unchanged are called **physical changes**. An example of a physical change is the conversion of ice into water. Both ice and water consist of molecules of H_2O (molecular composition stays the same) and the process does not involve changes to the nuclei of the various atoms (the H and O atoms within the molecules stay the same).

Processes that involve changes to molecular compositions but leave atomic identities unchanged are called chemical reactions. The burning of propane is an example of a chemical reaction as it involves changes in the groupings of atoms in propane, C_3H_8, and oxygen, O_2, to new groupings in carbon dioxide, CO_2, and water, H_2O. The various atoms themselves remain the same.

Chemical reactions often result from the regrouping of atoms that belong to the molecules of two or more substances that are mixed together. As an example, we can mix propane (the gas in cylinders used to heat food when we barbecue) and oxygen[18] to generate carbon dioxide and water. This is written as

$$C_3H_8 \; + \; O_2 \; \rightleftharpoons \; CO_2 \; + \; H_2O$$

where C_3H_8 is the molecular formula for propane. Depending on which direction is followed, the entities on one side of the double arrow are called **reactants** while those on the other side of the double arrow are called **products**. In the example above, going from left to right, the reactants are propane, C_3H_8, and oxygen, O_2, and the products are carbon dioxide, CO_2, and water, H_2O. As you can see the grouping of the atoms in the molecules has changed: We started with groupings of 3 C atoms and 8 H atoms within propane molecules as well as groupings of 2 O atoms within oxygen molecules and we ended up with groupings of 1 C atom and 2 O atoms within carbon dioxide molecules and 2 H atoms and 1 O atom within water molecules.

Given that, by agreement, chemical reactions result in regrouping of atoms only, such reactions can not result in changes to the atoms themselves which in turn implies that the total number of each type of atom before the reaction must equal to the total number of that kind of atom after the reaction. In the example above, the total number of C atoms before the reaction must equal to the total number of C atoms after the reaction. The same can be said about H atoms and O atoms: The total number of H atoms before the reaction must equal to the total number of H atoms after the reaction, and the total number of O atoms before the reaction must equal to the total number of O atoms after the reaction. However, when we look at the statement

$$C_3H_8 \; + \; O_2 \; \rightleftharpoons \; CO_2 \; + \; H_2O$$

it seems as if atoms have changed: We started with 3 C atoms and we ended up with 1 C atom which seems to imply that 2 C atoms disappeared during the process. Similarly we started with 8 H atoms but we ended up with 2 H atoms which seems to imply that 6 H atoms disappeared during the process. And while we started with 2 O atoms, we ended up with 3 O atoms which seems to imply that an extra O atom appeared at some point during the process.

Processes that involve changes in the nuclei of atoms are called **nuclear reactions**. Nuclear reactions often, though not always, involve changes to the identities of atoms themselves. An example of a nuclear reaction is the conversion of iodine-131, a radioisotope of iodine used in the diagnosis and treatment of medical issues related to the goiter, to xenon.

[18]By agreement, the word *oxygen* refers to the molecule O_2 and not an atom of oxygen. If one wishes to talk about an *atom* of oxygen, then one must refer to it as such. The same comment is true for hydrogen, H_2, nitrogen, N_2, and the halogens, F_2, Cl_2, I_2, and Br_2.

While reactions in which atoms are generated or destroyed *are* possible, such processes are properly classified as *nuclear reactions*, not chemical reactions. This means that, if we are certain, on other grounds, that the process of burning propane *is*, in fact, a chemical reaction, then it is not the case that a single molecule of C_3H_8 combines with a single molecule of O_2, generating a single molecule of CO_2 and a single molecule of H_2O. Rather, a number of each of these molecules are involved in the process. And if we try to figure out how many of each of the reactants and how many of each of the products will ensure that we can account for each and every C, H, and O atom after the reaction, we find that the following works:[19]

$$C_3H_8 \; + \; 5O_2 \quad \rightleftharpoons \quad 3CO_2 \; + \; 4H_2O$$

i.e., 1 molecule of C_3H_8 combines with 5 molecules of O_2, generating 3 molecules of CO_2 and 4 molecules of H_2O. The numbers 1, 5, 3 and 4 in this reaction are called the **stoichiometric coefficients** of C_3H_8, O_2, CO_2, and H_2O respectively. We refer to the reaction above as a **balanced reaction** as it ensures that the total number of each kind of atom before the reaction is equal to (balances) the total number of that kind of atom after the reaction. As an example, the balanced reaction for the burning of propane above tells us that starting with 3 C atoms (from 1 molecule of C_3H_8 which has 3 C atoms), we end up with 3 C atoms (from 3 molecules of CO_2 each of which has 1 C atom); starting with 8 H atoms (from 1 molecule of C_3H_8 which has 8 H atoms), we end up with 8 H atoms (from 4 molecules of H_2O each of which has 2 H atoms); and starting with 10 O atoms (from 5 molecules of O_2 each of which has 2 O atoms), we end up with 10 O atoms (from 3 molecules of CO_2 each of which has 2 O atoms, and 4 molecules of H_2O each of which has 1 O atom).

Stoichiometric coefficients in a chemical reaction are related through proportion: Twice the amount of any substance implies twice the amount of any other substance, and three times the amount of any substance implies three times the amount of any other substance, and so on.

A chemical reaction often involves heat. Heat is either supplied to the reaction (heating up the reactants to get them to react) or it is generated as a product (the mixture heats up as the reaction proceeds). If the heat term is of interest, it can be included in the chemical reaction as a term. As an example, combustion of 1 molecule of propane generates 23.02 eV of heat where eV is the unit symbol for the unit *electronvolt*, a tiny unit of energy suitable for working with energies at the level of atoms and molecules. We

---

[19]The reaction above is short for

$$1C_3H_8 \; + \; 5O_2 \quad \rightleftharpoons \quad 3CO_2 \; + \; 4H_2O$$

with 1 on the left of C_3H_8 omitted by convention.

write this as

$$C_3H_8 + 5O_2 \rightleftharpoons 3CO_2 + 4H_2O + 23.02 \text{ eV}$$

Note that the heat term is also related to the stoichiometric coefficients through proportion: Twice the amount of each of the reactants and products implies twice as much heat being used or generated, and three times the amount of each of the reactants and products implies three times as much heat being used or generated, and so on.

Exercise Set B.4

1. What is a physical process? A chemical reaction? A nuclear reaction?
2. Can atoms regroup during a physical process? A chemical reaction? A nuclear reaction?
3. Can atoms be generated or destroyed during a physical process? A chemical reaction? A nuclear reaction?
4. In each case identify the reactants and the products going from left to right.

    a. $CH_4 + O_2 \rightleftharpoons CO_2 + H_2O$
    b. $CO_2 + H_2O \rightleftharpoons C_6H_{12}O_6 + O_2$
    c. $Fe + S_8 \rightleftharpoons FeS$
    d. $Mg + H_2O \rightleftharpoons Mg(OH)_2 + H_2$
    e. $Fe + NaBr \rightleftharpoons FeBr_3 + Na$
    f. $NaOH + KNO_3 \rightleftharpoons NaNO_3 + KOH$
    g. $CaCl_2 + AgNO_3 \rightleftharpoons AgCl + Ca(NO_3)_2$
    h. $H_2SO_4 + KOH \rightleftharpoons K_2SO_4 + H_2O$

5. Which of the following reactions are correctly balanced?

    a. $CH_4 + 2O_2 \rightleftharpoons CO_2 + 2H_2O$
    b. $6CO_2 + 6H_2O \rightleftharpoons C_6H_{12}O_6 + 6O_2$
    c. $Fe + S_8 \rightleftharpoons 2FeS$
    d. $Mg + 3H_2O \rightleftharpoons Mg(OH)_2 + 2H_2$
    e. $Fe + 3NaBr \rightleftharpoons FeBr_3 + 3Na$
    f. $NaOH + KNO_3 \rightleftharpoons NaNO_3 + KOH$
    g. $CaCl_2 + 2AgNO_3 \rightleftharpoons AgCl + 3Ca(NO_3)_2$
    h. $H_2SO_4 + 2KOH \rightleftharpoons K_2SO_4 + 2H_2O$

B.5   Units of Measure

Two sets of units are needed for working with chemical substances: One consisting of units that are small enough to be suitable for working with individual atoms and molecules, and another, consisting of units that are large enough to be suitable for working with chunks of matter that we deal with on a daily basis. We limit the discussion to units of mass, energy, and amount.[20]

Let us begin with mass. The unit *atomic mass unit* is an extremely small unit of mass. As such, it is a suitable unit for the measurement of the values of the masses of individual atoms or molecules but not the values of the masses of chunks of substances that we deal with on a daily basis. The unit *gram*, on the other hand, is a suitable unit for the measurement of the values of the masses of chunks of substances that we work with on a daily basis but not the values of the masses of individual atoms or molecules.[21]

Let us now consider energy. The unit *electronvolt* is an extremely small unit of energy. As such, it is a suitable unit for the measurement of the values of energies that involve few atoms or molecules but not the values of energies that involve chunks of substances that we deal with on a daily basis. The unit *joule*, on the other hand, is a suitable unit for the measurement of the values of energies that involve chunks of substances that we work with on a daily basis but not values of energies that involve few atoms or molecules.[22]

We now consider amount. The unit *one* is a suitable unit for the measurement of the values of the amounts of atoms or molecules when we count a few of them but not for the measurement of the values of the amounts of atoms or molecules in chunks of substances that we work with on a daily basis. What we now need is a unit of measure for the quantity *amount* that is suitable for the measurement of the values of the amounts of atoms or molecules in chunks of substances that we work with on a daily basis.

Let us call this unit, whatever it may be, by the name *mole* with the corresponding unit symbol mol. A bit of reflection should convince the reader that the mole will have to be a collection of many entities.[23,24] In addition,

---

[20]We remind the reader that the quantity name *amount* is being used in its technical sense which refers to the number of entities being counted. We trust that the reader will keep this definition in mind whenever s/he encounters the quantity name *amount* and will not remind the reader of this interpretation of the quantity name *amount* in the pages ahead.

[21]$1 \text{ g} \cong 6.022 \times 10^{23}$ amu

[22]$1 \text{ J} \cong 6.242 \times 10^{18}$ eV

[23]In the same way that the unit of counting called *dozen* refers to a collection of 12 entities.

[24]For the reader who may not be quite sure what we mean by *bit of reflection* we provide an analogy: Consider the units of length *millimetre* and *kilometre*. The unit *millimetre* is a small unit of length, suitable for the measurement of the values of lengths that are small.

given that we want to use the mole to work with the huge amounts of atoms or molecules that are present in chunks of substances that we work with on a daily basis, the mole should refer to a collection of a huge number of entities. The question, now, is, what specific value should be associated with the unit mole, i.e., the collection of *exactly how many* entities should be referred to as 1 mol of that entity?

To answer this question, consider a carbon atom with a mass of 12.01 amu. Now consider this: If we keep the numerical value of the mass of a carbon atom but change its unit to grams, we arrive at 12.01 g, a value of mass that falls within the range of values of masses that correspond to chunks of matter that we work with on a daily basis. And if a mass of 12.01 g of carbon corresponds to a chunk of carbon that we work with on a daily basis, then the number of atoms of carbon in that chunk must also fall within the range of the number of carbon atoms that we work with on a daily basis. We, therefore, choose the number of atoms of carbon in 12.01 g of carbon as 1 mol of carbon.[25],[26]

In the argument above we chose 12.01 g of carbon as representative of chunks of carbon that we work with on a daily basis and set 1 mol of carbon atoms equal to the number of carbon atoms within the 12.01 g chunk. The reason we chose to call the number of atoms in 12.01 g of carbon (instead of 10 g of carbon or 15 g of carbon or some other reasonable value of mass of carbon in grams) is that this choice ensures that the numerical value of the mass of 1 mol of carbon atoms in grams and the numerical value of the mass of 1 carbon atom in amu are identical (they are both 12.01), allowing us to interpret the quoted numerical value of mass of carbon in the Periodic Table, i.e., 12.01, as either the atomic mass of carbon, i.e., 12.01 amu/atom, or the

---

The unit *millimetre*, however, is not suitable for the measurement of the values of large distances, such as distances between cities. A larger unit is needed to work with distances that are large. In choosing a larger unit of length, we can piece together many *millimetres* and refer to the sum of the lengths of these small units of length as a *kilometre*.

The relationship between the units of counting *one* and *mole* is similar to the relationship between the units of length *millimetre* and *kilometre*. The unit *one* is suitable for the measurement of the amounts of entities when there are few of them but not when we need to count a large number of entities. For measuring the amounts of entities when there is a large number of them, we need a larger unit of counting. A bit of reflection shows that just as we pieced together many small units of length (the many millimetres) to generate a larger unit of length (e.g., the *kilometre*), we can do the same with amount: We can group together many small units of amount (e.g., many *ones*) to generate a larger unit of amount as the collection of the individual *ones* (e.g., the *mole*). In other words, just as the net length of many *millimetres* generates the larger unit of length called *kilometre*, the net count of many *ones* generates the larger unit of counting called *mole*.

[25] 1 mol $\cong$ 6.022 $\times$ 10^{23} 1

[26] At this point we wish to point out that the official definition of mole refers to the number of carbon atoms in 12 g of ^{12}C and not 12.01 g of carbon atoms from a sample from nature which contains 99% ^{12}C and 1% ^{13}C. The argument given, of course, holds no matter what specific atom or isotope is chosen. The official definition is preferred experimentally as it imposes uniformity on the sample being used.

molar mass[27] of carbon, i.e., 12.01 g/mol.

This equality in the numerical values of the atomic mass and molar mass is not limited to carbon – The same argument given above holds no matter what specific atom is chosen. And the equality presents a significant advantage over other alternatives as its adoption implies that any calculations performed on mass may be interpreted as being in units of either amu/atom (if one needs to deal with a few atoms or molecules) or g/mol (if one needs to deal with chunks of matter).

Note also that in chemical reactions the stoichiometric coefficients may be interpreted in the sense of the mol as well. In

$$C_3H_8 + 5O_2 \rightleftharpoons 3CO_2 + 4H_2O$$

we can say that 1 molecule of C_3H_8 reacts with 5 molecules of O_2, generating 3 molecules of CO_2 and 4 molecules of H_2O. We can also say that 1 mol of C_3H_8 reacts with 5 mol of O_2, generating 3 mol of CO_2 and 4 mol of H_2O. This dual interpretation results from the fact that the stoichiometric coefficients in chemical reactions are directly proportional to each other; a claim that we justified earlier in this appendix.

If the heat term is present, then the interpretation of the unit of the stoichiometric coefficients may be limited to either *one* or the *mole*. The stoichiometric coefficients in the reaction

$$C_3H_8 + 5O_2 \rightleftharpoons 3CO_2 + 4H_2O + 23.02 \text{ eV}$$

are properly interpreted as having the unit *one* as the heat term is extremely small and, therefore, it is likely that the heat is generated by the combustion of a single molecule of propane. The stoichiometric coefficients in the reaction

$$C_3H_8 + 5O_2 \rightleftharpoons 3CO_2 + 4H_2O + 2220 \text{ kJ}$$

on the other hand, should be interpreted in the sense of the *mole* as the heat term is rather large and, therefore, it is unlikely that the heat is generated by the combustion of a single molecule of propane.

Exercise Set B.5

1. a. Name a unit of mass that is suitable for measuring the values of the masses of small amounts of atoms or molecules. Explain why this is a suitable unit of mass for measuring the values of the masses of small amounts of atoms or molecules. Write the corresponding unit symbol.

---

[27]The *molar mass* of an entity is the mass per 1 mol of that entity.

b. Name a unit of energy that is suitable for measuring the values of the energies of small amounts of atoms or molecules. Explain why this is a suitable unit of energy for measuring the values of the energies of small amounts of atoms or molecules. Write the corresponding unit symbol.

c. Name a unit of amount that is suitable for measuring the values of the amounts of small amounts of atoms or molecules. Explain why this is a suitable unit of amount for measuring the values of the amounts of small amounts of atoms or molecules. Write the corresponding unit symbol.

2. a. Name a unit of mass that is suitable for measuring the values of the masses of chunks of matter. Explain why this is a suitable unit of mass for measuring the values of the masses of chunks of matter. Write the corresponding unit symbol.

b. Name a unit of energy that is suitable for measuring the values of the energies of chunks of matter. Explain why this is a suitable unit of energy for measuring the values of the energies of chunks of matter. Write the corresponding unit symbol.

c. Name a unit of amount that is suitable for measuring the values of the amounts of chunks of matter. Explain why this is a suitable unit of amount for measuring the values of the amounts of chunks of matter. Write the corresponding unit symbol.

3. Why do we refer to the number of atoms in 12.01 g of carbon as 1 mol of carbon? Why not call the number of atoms in, say, 10 g of carbon or 15 g of carbon as 1 mol of carbon?

4. What is atomic mass? How is atomic mass computed?

5. What is molar mass? How is molar mass computed?

6. Using the Periodic Table in Figure B.2.2 determine the molar mass of each of the following atoms.

a. O	c. Tc	e. F	g. U
b. S	d. I	f. H	h. P

7. In each case, calculate the total number of each kind of atom. Interpret the number on the left in the sense of the mole.

a. $5C_2H_6$	c. $4S_8$	e. $NaNO_3$	g. $6H_2SO_4$
b. $12CO_2$	d. $3H_2O$	f. $15CaCl_2$	h. CO

Appendix C
A Primer on Grading Schemes

In this appendix we will discuss three different grading schemes and the relationships between them. These are percent grades, letter grades and grade points.

C.1 Grading Schemes

Percent grades, letter grades and grade points represent three common grading schemes used to communicate a student's grades at educational institutions.

Percent grades are the most familiar of the grading schemes. They are based on a scale that runs from 0% to 100%. The scale used is often discrete with grades rounded to the nearest percent.

Letter grades employ the letters A, B, C, D and F, either on their own or along with a + or a − as postfix, except for F which can only appear on its own. Examples are B, C+, D−, and F but not F− or F+.

Grade points use a discrete scale that runs from 0.00 to 4.33.[1] Possible values on this scale are 0.00, 0.67, 1.00, 1.33, 1.67, 2.00, 2.33, 2.67, 3.00, 3.33, 3.67, 4.00 and 4.33. We will discuss the logic behind these particular grade point values below.

C.2 Relationships between Grading Schemes

The relationship between percent grades, letter grades and grade points is shown in Table C.2.1 below:[2]

---

[1]Some educational institutions use a scale that runs from 0.00 to 4.00.

[2]The relationships shown in this table are standard relationships but the reader should note that variations exist. As an example, as noted above, some educational institutions

Percent Grades	Letter Grades	Grade Points
90%+	A+	4.33
86%–89%	A	4.00
80%–85%	A–	3.67
77%–79%	B+	3.33
73%–76%	B	3.00
70%–72%	B–	2.67
67%–69%	C+	2.33
63%–66%	C	2.00
60%–62%	C–	1.67
57%–59%	D+	1.33
53%–56%	D	1.00
50%–52%	D–	0.67
49%–	F	0.00

Table C.2.1: Relationship between percent grades,
letter grades and grade points

Let us take a closer look at the relationships between the different grading schemes in Table C.2.1 above.

Relationship Between Percent Grades and Letter Grades

We will start with letter grade F and work our way up.

First note that any percent grade below 50% is equivalent to an F.

Next, the letters D, C and B are associated with the percent grades in the 50% range, 60% range and 70% range respectively.

Within each of these percent ranges, the lower third is represented by a – sign written as a postfix to the letter grade, the middle third is represented by the letter grade itself and the top third is represented by a + sign written as a postfix to the letter grade. As an example, the letter grade C– corresponds to the lower third of the 60% range, i.e., from 60% to 62% (inclusive on both ends), the letter grade C corresponds to the middle third of the 60%

---

use a grade point scale that runs from 0.00 to 4.00 and some associate a grade point value of 1.00 to the letter grade D– as well as D. The reader should always contact the institution they attend to obtain a copy of the relationships between the various grading schemes and use these in matters that relate to her/his grades.

range, i.e., from 63% to 66% (inclusive on both ends) and the letter grade C+ corresponds to the top third of the 60% range, i.e., from 67% to 69% (inclusive on both ends).[3]

The A range is associated with the 80%+ range. An A− covers the range 80% to 85% (inclusive on both ends), an A covers the range 86% to 89% (inclusive on both ends) and an A+ covers the range 90%+.

Relationship Between Percent Grades and Grade Points

The relationship between percent grades and grade points follows a pattern that is similar to the one between percent grades and letter grades.

First note that the percent grades below 50% are given a grade point value of 0.00.

Next, an association is made between percent grades in the 50% range and the grade point value 1.00 as follows. As before, the 50% range is subdivided into three parts: The lower third which covers the percent grades 60%, 61% and 62% is set to correspond to a grade point value of $1.00 - \frac{1}{3}$ which evaluates to 0.67.[4] The middle third which covers the percent grades 63%, 64%, 65% and 66% is set to correspond to a grade point value of 1.00, and the top third which covers the percent grades 67%, 68% and 69% is set to correspond to a grade point value of $1.00 + \frac{1}{3}$ which evaluates to 1.33.

Similar associations are made between percent grades in the 60% and 70% ranges and grade point values 2.00 and 3.00: The percent grades 60%, 61% and 62% are associated with the grade point value of $2.00 - \frac{1}{3}$ or 1.67, the percent grades 63%, 64%, 65% and 66% are associated with the grade point value of 2.00, and the percent grades 67%, 68% and 69% are associated with the grade point value of $2.00 + \frac{1}{3}$ or 2.33. Furthermore, the percent grades 70%, 71% and 72% are associated with the grade point value of $3.00 - \frac{1}{3}$ or 2.67, the percent grades 73%, 74%, 75% and 76% are associated with the grade point value of 3.00, and the percent grades 77%, 78% and 79% are associated with the grade point value of $3.00 + \frac{1}{3}$ or 3.33.

---

[3]Note that a C− covers the three percent grades 60%, 61% and 62%, a C covers the four percent grades 63%, 64%, 65% and 66%, and a C+ corresponds to the percent grades 67%, 68% and 69%. The reason a C covers a larger set of grades is due to rounding.

To explain how, note that there are 10 grades in total from 60% to 69%. Each letter grade C−, C and C+ should, therefore, be given $3\frac{1}{3}$ of the total. Converting this share of the total to a decimal yields $3.333\cdots$ which means that a C− should include grades that range from 60% to $62.333\cdots$%, a C+ should include grades that range from $66.666\cdot$% to 69%, and a C should include whatever is left. However, since percent grades are often rounded to the nearest percent, a C− would end up covering the grades 60%, 61% and 62%, a C+ would end up covering the grades 67%, 68% and 69%, and a C would cover the rest of the percent grades in the 60% range, i.e., 63%, 64%, 65% and 66%.

[4]Calculations involving grade points are rounded to two decimal digits.

In the 80%+ range, the percent grades from 80% to 85% (inclusive on both ends) are associated with the grade point value of $4.00 - \frac{1}{3}$ or 3.67, the percent grades from 86% to 89% (inclusive on both ends) are associated with the grade point value of 4.00 and the percent grades from 90% to 100% (inclusive on both ends) are associated with the grade point value of $4.00 + \frac{1}{3}$ or 4.33.[5]

Relationship Between Letter Grades and Grade Points

To explain how letter grades are related to grade points, we begin by assigning grade points to the main letters themselves. This is shown in Table C.2.2 below:

Letter Grades	Grade Points
A	4.00
B	3.00
C	2.00
D	1.00
F	0.00

Table C.2.2: Relationship between the main letter grades and grade points

As Table C.2.2 shows, the grade point values 0.00, 1.00, 2.00, 3.00 and 4.00 are associated with the letter grades F, D, C, B and A. Other than the F letter grade and its associated grade point value of 0.00, the rest of the intervals covered by letter grades and grade points are divided into three parts with corresponding parts being associated with each other. As an example, on the letter grade side, the C region is divided into the three regions C−, C and C+. On the grade point side, the 2.00 region is divided into the three regions $2.00 - \frac{1}{3}$ or 1.67, 2.00, and $2.00 + \frac{1}{3}$ or 2.33. We, therefore, associate the letter grade C− to grade point value 1.67, the letter grade C to grade point value 1.00 and the letter grade C+ to grade point value 2.33.

---

[5]Some institutions assign the grade point value 4.00 to all grades from 86% to 100% (inclusive on both ends).

Appendix D
Alternative Models for Problems of Type *Direct Proportion*

In this appendix we will discuss alternative models for problems of type *direct proportion*. Such alternatives may be classified as either extensions of the standard form or the conservation form of the equation for direct proportion.[1] Each type can be setup using rates or ratios.[2]

D.1 Standard Models for Problems of Type *Direct Proportion*

This approach was presented in the body of the textbook and is the approach that we recommend for use in modelling word problems. The approach is based on the standard form of the equation for direct proportion, i.e., $q_2 = kq_1$, and its extension, $q_2 = k_n \dots k_2 k_1 q_1$, with q_1 and q_2 representing the values of the two quantities involved and the various k, which are constants, representing the rates or ratios in the models.

As an example of how this approach works, consider the following problem.

---

[1] For a detailed discussion of the standard and conservation forms of the equation for direct proportion please see the companion textbook *Style in Technical Math* by the author.

[2] Recall that a *rate* is a comparison of the values of two quantities with different units by division. As an example of a rate, consider the problem of covering a distance of 240 km in 2 h. The average speed may then be computed as the rate $\frac{240 \text{ km}}{2 \text{ h}}$ which simplifies to 120 km/h. Note that a rate has a unit other than 1.

A *ratio*, on the other hand, is a comparison of the values of two quantities with the same unit. As an example of a ratio, if for every \$2 that I make, you make \$3, then the ratio of the money that you make to the money that I make is 3 : 2 (read 3 *to* 2) which may be written as $\frac{3}{2}$. In a ratio, the units that are used are identical and, since they are being divided, they generate the unit 1.

Problem

3 pens cost \$3.60. At this rate, what is the cost of 2 pens?

An algebraic model for this problem using rates is given below.

An Algebraic Model using Rates

c cost of the pens (\$)

$$c = \frac{3.60}{3} \times 2$$

As discussed in the body of the textbook, the model can be setup by noting the given relationship

3 pens   \$3.60

Next, we begin with the value that we want to process, i.e., 2 pens. We write

3 pens   \$3.60
2 pens

Since there is only one relationship listed, our model will have one division as shown below.

3 pens   \$3.60
——— $\times$ 2 pens

The values in the relationship above will appear as the dividend and divisor of this division with the one that has the same unit as 2 (i.e., 1 or, informally, *pens*) appearing as the divisor, i.e.,

3 pens   \$3.60

$$\frac{\$3.60}{3 \text{ pens}} \times 2 \text{ pens}$$

The logic behind this model lies in the conversion of the rate $\frac{\$3.60}{3 \text{ pens}}$ to the unit rate \$1.20/pen, and then multiplying this by 2 to find the cost of 2 pens.

One may also use ratios to set up a model for the problem above. This alternative is based on the logic that states that *the cost of 2 pens is $\frac{2}{3}$ of the cost of 3 pens*. This translates verbatim into a model.

An Algebraic Model using Ratios

c cost of the pens ($)

$$c = \frac{2}{3} \times 3.60$$

We now explain the manner in which the model above is set up. Following this we will present the reason why it works.

To setup the model, we begin by noting the given relationship

3 pens $3.60

Since there is one relationship, our model will have one division as shown below.

3 pens $3.60

$\overline{} \times$

We begin with the value that we want to process, i.e., 2 pens. We write

3 pens $3.60

$\dfrac{2 \text{ pens}}{} \times$

The values in the relationship above will appear as the divisor of this division and the factor that follows the multiplication with the one that has the same unit as 2 (i.e., 1 or, informally, *pens*) appearing as the divisor, i.e.,

3 pens $3.60

$\dfrac{2 \text{ pens}}{3 \text{ pens}} \times \$3.60$

As stated above, the logic behind this model states that the cost of 2 pens is $\frac{2}{3}$ of the cost of 3 pens.

In some problems, as in the one posed above, the use of either rates and ratios follows the kinds of logic that we naturally use. In some cases, however, the logic that uses rates or the one that uses ratios may be more in line with the logic that we normally use. Here is an example of a case where the use of rates has more immediate meaning.

Problem

Convert a length of 4.8 m to ft given that

1 m $=$ 100 cm
2.54 cm $=$ 1 in
12 in $=$ 1ft

An Algebraic Model using Rates

L length (ft)

$$L = \frac{1}{12} \times \frac{1}{2.54} \times \frac{100}{1} \times 4.8$$

As explained in the body of the text, this model is an extension of the one that has a single rate in its solution. We let the reader explain how the model above is put together.

We can see how this model works by working our way back from the right factor by factor. The rightmost factor, i.e.,

4.8 m

represents the value whose units we want to change. Moving one factor to the left, we have the subexpression

$$\frac{100 \text{ cm}}{1 \text{ m}} \times 4.8 \text{ m}$$

Converting the rate $\frac{100 \text{ cm}}{1 \text{ m}}$ yields the unit rate 100 cm/m, i.e., there are 100 cm in 1 m. Therefore, to find the length in cm, we multiply 4.8 m by 100 cm/m. This yields 480 cm. We now move one factor to the left and try to make sense of

$$\frac{1 \text{ in}}{2.54 \text{ cm}} \times 480 \text{ cm}$$

Again we convert the rate $\frac{1 \text{ in}}{2.54 \text{ cm}}$ to the unit rate 0.393 700 ... in/cm, i.e., there are 0.393 700 ... in in 1 cm. To find the length in in, we multiply the length in cm by 0.393 700 ... in/cm to get 188.976... in.

We let the reader explain the logic behind the last link, i.e.,

$$\frac{1 \text{ ft}}{12 \text{ in}} \times 188.976... \text{ in}$$

The logic behind the use of ratios to setup a model for this problem is not as straightforward. Here is the model.

An Algebraic Model using Ratios

L length (ft)

$$L = \frac{4.8}{1} \times \frac{100}{2.54} \times \frac{1}{12} \times 1$$

To set up the expression on the right side of the model above, we begin by noting the relationships that are given:

1 m = 100 cm
2.54 cm = 1 in
12 in = 1 ft

Since there are three relationships, our expression will involve three ratios as shown below:

1 m = 100 cm
2.54 cm = 1 in
12 in = 1 ft

$$\frac{}{} \times \frac{}{} \times \frac{}{}$$

We now begin with the value whose unit we wish to convert, i.e., 4.8 m, and we write:

1 m = 100 cm
2.54 cm = 1 in
12 in = 1 ft

$$\frac{4.8 \text{ m}}{} \times \frac{}{} \times \frac{}{}$$

The values in one of the relationships listed will form the divisor of the leftmost division and the dividend of the middle division with the value that has the same unit as that of 4.8, i.e., m, appearing as the divisor of the leftmost division. The only relationship that has m is 1 m = 100 cm. We write

1 m = 100 cm   ✓
2.54 cm = 1 in
12 in = 1 ft

$$\frac{4.8 \text{ m}}{1 \text{ m}} \times \frac{100 \text{ cm}}{} \times \frac{}{}$$

Note that we have checked off the relationship whose values were just used. This reminds us that the values in this relationship should no longer be used.

The values in one of the remaining relationships will appear as the divisor of the middle division and the dividend of the rightmost division with the value that has the same unit as the unit of 100, i.e., cm, appearing as the divisor of the middle division. This yields the following.

1 m = 100 cm   ✓
2.54 cm = 1 in   ✓
12 in = 1 ft

$$\frac{4.8 \text{ m}}{1 \text{ m}} \times \frac{100 \text{ cm}}{2.54 \text{ cm}} \times \frac{1 \text{ in}}{} \times$$

The values in the last relationship appear as the divisor of the rightmost division and the factor that follows the rightmost multiplication with the value that has the same unit as 1, i.e., in, appearing as the divisor of the rightmost division. We now have the following expression

$$1 \text{ m} = 100 \text{ cm} \quad \checkmark$$
$$2.54 \text{ cm} = 1 \text{ in} \quad \checkmark$$
$$12 \text{ in} = 1 \text{ ft} \quad \checkmark$$

$$\frac{4.8 \text{ m}}{1 \text{ m}} \times \frac{100 \text{ cm}}{2.54 \text{ cm}} \times \frac{1 \text{ in}}{12 \text{ in}} \times 1 \text{ ft}$$

Dropping the units leads to the following model.

$$L = \frac{4.8}{1} \times \frac{100}{2.54} \times \frac{1}{12} \times 1$$

While it is possible to associate meaning to this set up, the semantics behind the model are not as clear as the one that uses rates.[3]

The use of ratios is preferred when the problem at hand discusses *a fraction of a value, a decimal fraction of a value,* or *a percentage of a value.* Here is an example.

Problem

There are 120 frogs in the sample. $\frac{2}{3}$ of the frogs in the sample have red eyes. Of these, $\frac{1}{4}$ have green skin. How many green-skinned, red-eyed frogs are there in the sample?

Here is the model using ratios.

$$n = \frac{1}{4} \times \frac{2}{3} \times 120$$

where n represents the number of green-skinned, red-eyed frogs in the sample. The model states that *the number of green-skinned, red-eyed frogs in the sample is equal to* $\frac{1}{4}$ *of* $\frac{2}{3}$ *of* 120.

The semantics behind the model that uses rates is not so clear:

$$\frac{120}{3} \times \frac{2}{4} \times 1$$

The standard models given above may be written using a single division notation. As an example, the model using rates for the conversion of unit of

---

[3] We invite the reader to try to work out the semantics behind the model, starting with the factor on the right and moving left factor by factor.

length in 4.8 m given above, i.e.,

$$L = \frac{1}{12} \times \frac{1}{2.54} \times \frac{100}{1} \times 4.8$$

may be written as

$$L = \frac{1 \times 1 \times 100 \times 4.8}{12 \times 2.54 \times 1}$$

with the factors that are 1 often dropped to generate

$$L = \frac{100 \times 4.8}{12 \times 2.54}$$

Such models may be setup as paraphrases of the ones above that use rates and ratios or else they can be worked out by noting whether each factor in the dividend and divisor of the division on the right is directly or inversely proportional to the value on the left. Many science formulas use this setup in their models.

D.2 Conservation Models for Problems of Type *Direct Proportion*

The conservation form of direct proportion is based on the argument that, if two quantities are related through direct proportion, then their rates or ratios are constant. This should make sense: If the two quantities involved are related through direct proportion, then doubling the value of one of them doubles the value of the other; tripling the value of one of them triples the value of the other, and so on. But if both values double, triple, etc., then their rates or ratios remain the same. As an example, suppose q_2 is directly proportional to q_1 and that when q_1 is 5, q_2 is 15. This leads to the ratio of $\frac{15}{5}$ or 3 for the ratio $\frac{q_2}{q_1}$. If we double q_1 and q_2 to 10 and 30, the ratio $\frac{q_2}{q_1}$, or $\frac{30}{10}$, remains the same: 3. Try tripling q_1 and q_2 and see what happens to their ratio. This constancy of the rates or ratios provides us with the familiar model for direct proportion problems as equality of two rates or ratios.

As an example of the conservation form of the model for direct proportion problems, consider the problem posed earlier, reproduced below.

Problem

3 pens cost \$3.60. At this rate, what is the cost of 2 pens?

Here is the model following the conservation form using rates.

An Algebraic Model using Rates

$$\frac{3.60}{3} = \frac{c}{2}$$

And the following is the model following the conservation form using ratios.

An Algebraic Model using Ratios

$$\frac{2}{3} = \frac{c}{3.60}$$

In both models c represents the cost of 2 pens.

We now explain how the model is put together. Following this, we will present the reason why it works.

To explain how conservation models for direct proportion problems are set up, we begin by identifying the two quantities involved in the problem. In the case of the problem above, these are the number of pens purchased and the cost of the pens.

Next, we note that all conservation models have the following form.

$$\underline{} = \underline{}$$

Following this, we must make a decision as to whether we wish to set up the model using rates or ratios. We will discuss each in turn.

To set up the model using rates, we associate one of the quantities involved (say, the cost of the pens) to the dividends of the two ratios. Associating the cost of the pens to the dividends in the problem above, this yields the following.

$$\frac{3.60}{} = \frac{c}{}$$

where we have written the cost of the 3 pens, i.e., \$3.60, as the dividend of the division on the left, and the cost of the 2 pens, i.e., the unknown c, as the dividend of the division on the right.

Next, we associate the other quantity, i.e., the number of pens purchased in the problem above, to the divisors of the two divisions, making sure that the order in which they are placed matches the order in which their corresponding costs are placed. This yields the following:[4]

$$\frac{3.60}{3} = \frac{c}{2}$$

Having explained how to set the model up using the conservation formulation, we now present the reason why it works.

The logic behind this formulation is quite simple: The rates on the left and the right side of the model both calculate the cost of a pen: The expression on the left, i.e., $\frac{3.60}{3}$, does so by dividing the cost of 3 pens by 3 and the expression on the right, i.e., $\frac{c}{2}$, does the same by dividing the cost of 2 pens (albeit unknown) by 2. Since both expressions compute the cost of a pen, they must be equal to each other.

We let the reader present an algorithm for setting up the conservation model that uses ratios and the present the reason why it works.

The manner in which conservation models are solved is discussed in the companion textbook *Style in Technical Math* by the author.

Conservation formulation of direct proportion problems provide an alternative to the standard model in modelling single-proportion problems.

The standard model is generally preferred and is the model of choice when the rates or ratios involved have significance. An example of this is the equation for motion along a straight line at constant speed, i.e., $d = vt$. This equation relates the distance covered, d, to travel time, t, through multiplication of t by v, i.e., speed which is a rate.

The conservation model is generally preferred when the relevant rate or ratio does not have physical significance[5] and where one wishes to bring attention to the equality of the relevant rates or ratios. As an example, the direct proportion relationship between the pressure of a gas in a container in

---

[4]Alternative formulations are possible. As an example, we might choose to list the values that relate 2 pens to their cost first. This yields the following model.

$$\frac{c}{2} = \frac{3.60}{3}$$

Or we may choose to associate the number of pens purchased to the dividends and the cost of the pens to the divisors. This generates either

$$\frac{3}{3.60} = \frac{2}{c}$$

or

$$\frac{2}{c} = \frac{3}{3.60}$$

[5]In the way that speed had significance in the equation $d = vt$.

Pa and its amount in mol is often formulated as

$$\frac{p_1}{p_2} = \frac{n_1}{n_2}$$

The main disadvantage of the conservation models for direct proportion problems is that their extension to modelling problems that involve multiple proportions is not simple and attempts to do so leads to the generation of systems. In spite of this major shortcoming, we do recommend that the reader study the manner in which conservation models are set up and solved as the model is used extensively by the average person as well as in the formulation of certain formulas in the sciences.

Appendix E
Alternative Models for Problems of Type *Inverse Proportion*

In this appendix we will discuss alternative models for problems of type *inverse proportion*. Such alternatives may be classified as either extensions of the standard form or the conservation form of the equation for inverse proportion.[1]

E.1 Standard Models for Problems of Type *Inverse Proportion*

This approach was presented in the body of the textbook and is the approach that we recommend for use in modelling word problems. The approach is based on the standard form of the equation for inverse proportion, i.e., $q_2 = \frac{k}{q_1}$ with k representing a constant.

As an example of how this approach works, consider the following problem.

Problem

8 painters can paint an office in 3.5 h. At this rate, how many painters are needed to paint the office in 2 h?

An algebraic model for this problem using rates is given below.

---

[1]For a detailed discussion of the standard and conservation forms of the equation for inverse proportion please see the companion textbook *Style in Technical Math* by the author.

An Algebraic Model

n number of painters needed (1)

$$n = \frac{8 \times 3.5}{2}$$

As discussed in the body of the textbook, the model can be setup by noting the given relationship

8 painters   3.5 h

Next, we note that, since this is an inverse proportion relationship, the model will have the form of a division. We write

8 painters   3.5 h

———

The related values in the relationship above will appear as factors in the dividend of this division, i.e.,

8 painters   3.5 h
8×3.5
———

The value that we wish to process, i.e., 2 h, appears as the divisor of this division, i.e.,

8 painters   3.5 h
$$\frac{8 \times 3.5}{2}$$

The key to understanding the model above lies in the understanding that, in problems of type inverse proportion, products of pairs of related values, such as the pair of related values 8 painters and 3.5 h in the problem above, are constant. This means that the product of the unknown number of painters and the desired duration of 2 h must also be equal to this constant. Therefore, the unknown number of painters must be the quotient of this constant, which may be computed as the product of 8 painters and 3.5 h, and 2 h.

The main disadvantage of this formulation is that its extension to modelling problems that involve multiple proportions is not simple and attempts to do so leads to the generation of complex fractions. An alternative approach is to use ratios to set up the model for inverse proportion problems. This model works better in cases where we need to set up a chain inverse proportion problem.

Following this alternative, we model the problem above as

An Algebraic Model

n number of painters needed (1)

$$n = \frac{3.5}{2} \times 8$$

To see how this model is set up, we begin with the given relationship, i.e.,

8 painters   3.5 h

Next, we begin with the value whose unit is the same as the unit of the quantity whose value we seek, i.e., 8 painters.

8 painters   3.5 h

8 painters

Since there is one relationship, the model will involve one division. This is shown below.

8 painters   3.5 h

—— × 8 painters

We now place the value 3.5 h as the dividend of this division. Note that 3.5 h is the value that, according to the relationship listed, corresponds to the value 8 painters.

8 painters   3.5 h

$\dfrac{3.5 \text{ h}}{\quad}$ × 8 painters

The value 2 h will appear as the divisor of this division.

8 painters   3.5 h

$\dfrac{3.5 \text{ h}}{2 \text{ h}}$ × 8 painters

We let the reader work out the logic behind this model.

E.2   Conservation Models for Problems of Type *Inverse Proportion*

The conservation form of inverse proportion is based on the argument that, if two quantities are related through inverse proportion, then their product is constant. This should make sense: If the two quantities involved are related through inverse proportion, then doubling the value of one of them halves the value of the other; tripling the value of one of them reduces the value of the other to one third of its value, and so on. But if one value doubles and the other halves, or one value triples while the other reduces to one third of its value, etc., then their products remain the same. As an example, suppose q_2 is inversely proportional to q_1 and that when q_1 is 3, q_2 is 8. This leads to a product of 8×3 or 24 for the product $q_2 q_1$. If we double q_1 and halve q_2 to 6 and 4, the product $q_2 q_1$, or 4×6, remains the same: 24. Try tripling q_1 and dividing q_2 by 3 and see what happens to their product. This constancy of the products provides us with the familiar model for inverse proportion problems as equality of two products.

As an example of the conservation form of the model for inverse proportion problems, consider the problem posed earlier, reproduced below.

Problem

> 8 painters can paint an office in 3.5 h. At this rate, how many painters are needed to paint the office in 2 h?

Here is the model following the conservation form using rates.

An Algebraic Model using Rates

$$2 \times n = 8 \times 3.5$$

where n represents the number of painters needed.

The logic behind this formulation is quite simple: The products on the left and the right side of the model both calculate the total amount of human·hours that are needed to complete the job. This value is constant and it can be computed through the expression 8×3 or $2 \times n$.

The manner in which conservation models are solved is discussed in the companion textbook *Style in Technical Math* by the author.

Conservation formulation of inverse proportion problems provide an alternative to the standard model in modelling single-proportion problems.

The main disadvantage of the conservation models for inverse proportion problems is that their extension to modelling problems that involve multiple proportions is not simple and attempts to do so leads to the generation of systems. In spite of this major shortcoming, we do recommend that the reader study the manner in which conservation models are set up and solved as the model is used extensively by the average person as well as in the formulation of certain formulas in the sciences. An example of such use it in the formula $p_1 V_1 = p_2 V_2$ in thermodynamics.

Index

reactants, 297

sink, 72
solution, **8**
solving a word problem, **8**
source, 72
square brackets, 34
step-by-step BEDMAS algorithm,
        18
step-by-step technique, **8**
stoichiometric coefficients, 298
subatomic particles, 284
subexpression, 65, 110
superposition, 59

term, **15**

unit, 278
unit name, 278
unit rate, 122
unit symbol, 279
US Customary System of Units, 275

value of a quantity, 280
value of quantity, 277

weight, 47
weighted average, 47
word problem, **3**

www.ingramcontent.com/pod-product-compliance
Lightning Source LLC
Chambersburg PA
CBHW060322200326
41519CB00011BA/1810